Emerging Chemicals and Human Health

Yunhui Zhang
Editor

Emerging Chemicals and Human Health

Springer

Editor
Yunhui Zhang
Key Laboratory of Health Technology
Assessment, National Health
Commission of the People's
Republic of China
Fudan University
Shanghai, China

Key Laboratory of Public Health
Safety, Ministry of Education
Fudan University
Shanghai, China

ISBN 978-981-32-9537-7 ISBN 978-981-32-9535-3 (eBook)
https://doi.org/10.1007/978-981-32-9535-3

This Springer imprint is published by the registered company Springer Nature Singapore Pte Ltd.
The registered company address is: 152 Beach Road, #21-01/04 Gateway East, Singapore 189721, Singapore

Contents

Contributors

Shangqin Chen Neonatal Intensive Care Unit, The Second Affiliated Hospital and Yuying Children's Hospital of Wenzhou Medical University, Wenzhou, China

Haixia Dai State Environmental Protection Key Laboratory of Formation and Prevention of Urban Air Pollution Complex, Shanghai Academy of Environmental Sciences, Shanghai, China

Renshan Ge Institute of Reproductive Biomedical Research, The Second Affiliated Hospital and Yuying Children's Hospital of Wenzhou Medical University, Wenzhou, China

Yuting Jin The Second Affiliated Hospital and Yuying Children's Hospital of Wenzhou Medical University, Wenzhou, China

Jialin Li School of Public Health, Fudan University, Shanghai, China
Key Laboratory of Public Health Safety, Ministry of Education, Fudan University, Shanghai, China

Cuiqing Liu College of Public Health, Zhejiang Chinese Medical University, Hangzhou, China

Ranran Luo School of Public Health, Fudan University, Shanghai, China
Key Laboratory of Public Health Safety, Ministry of Education, Fudan University, Shanghai, China

Wenjuan Ma MOE-Shanghai Key Laboratory of Children's Environmental Health, Xinhua Hospital, Shanghai Jiao Tong University School of Medicine, Shanghai, China

Junyi Ni Shanghai Experimental School, Shanghai, China

Pengpeng Wang School of Public Health, Fudan University, Shanghai, China
Key Laboratory of Public Health Safety, Ministry of Education, Fudan University, Shanghai, China

Chunhua Wu School of Public Health, Fudan University, Shanghai, China
Key Laboratory of Public Health Safety, Ministry of Education, Fudan University, Shanghai, China

Bin Xia School of Public Health, Fudan University, Shanghai, China
Key Laboratory of Health Technology Assessment, National Health Commission of the People's Republic of China, Fudan University, Shanghai, China

Xirong Xiao Department of Obstetrics and Gynecology, Obstetrics and Gynecology Hospital of Fudan University, Shanghai, China

Yunhui Zhang Key Laboratory of Health Technology Assessment, National Health Commission of the People's Republic of China, Fudan University, Shanghai, China
Key Laboratory of Public Health Safety, Ministry of Education, Fudan University, Shanghai, China

Yan Zhao Shanghai First Maternity and Infant Hospital, Tongji University School of Medicine, Shanghai, China

Yingya Zhao School of Public Health, Fudan University, Shanghai, China
Key Laboratory of Public Health Safety, Ministry of Education, Fudan University, Shanghai, China

Yuhan Zhou School of Public Health, Fudan University, Shanghai, China
Key Laboratory of Public Health Safety, Ministry of Education, Fudan University, Shanghai, China

Qingyang Zhu School of Public Health, Fudan University, Shanghai, China
Key Laboratory of Public Health Safety, Ministry of Education, Fudan University, Shanghai, China

Introduction: Emerging Chemicals of Concern

Yunhui Zhang

Abstract Emerging chemicals widely exist in the environment, potentially harmful to the ecological environment and human health, and also the preferred substances for future environmental control. This chapter provides a brief overview of the definition and classification of emerging chemicals and the screening scheme for those substances. Moreover, a list of the currently and most frequently discussed emerging chemicals is given, their effects on environment and human health as well as the sensitive population are discussed. Some protective tips are recommended at the end of this chapter.

Keywords Emerging chemical · Health effect · Children

Man-made synthetic chemicals have become an integral part of modern human life. These chemicals bring us not only a lot of convenience but also the environmental load and health risks. Nowadays, there are about 80,000 kinds of synthetic chemicals worldwide. However, a very little information about the complete toxicity of environmental toxicants is known to the public or academia. For instance, only 7% of 2863 high-yield industrial chemicals (total production and import amounts to more than 1 million tons) have data of hazard assessment, including acute toxicity, ecological toxicity, environmental fate, chronic toxicity, mutagenicity, and ultimately, reproductive and developmental toxicity. According to the reports of Hartung et al., 45% of high-volume chemicals available on the market have no basic toxicity information on either human health or environmental impact. The main reason for this data gap in health-based risk assessment is the involvement of huge cost in the risk assessment of chemicals. In fact, only about 5% of high-volume chemicals pass through the assessment and have a full set of basic toxicity information [1–3].

Y. Zhang (✉)
Key Laboratory of Health Technology Assessment, National Health Commission of the People's Republic of China, Fudan University, Shanghai, China

Key Laboratory of Public Health Safety, Ministry of Education, Fudan University, Shanghai, China
e-mail: yhzhang@shmu.edu.cn

Y. Zhang (ed.), *Emerging Chemicals and Human Health*,
https://doi.org/10.1007/978-981-32-9535-3_1

In the past decade, a great deal of such high volume chemical compounds have been added to the Persistent Organic Pollutants (POPs) list of the Stockholm Convention, for example flame retardants (FRs) and per-/polyfluoroalkyl substances (PFASs), organochloride pesticides (OCPs), polychlorinated biphenyls (PCBs), dechlorane plus (DPs). These chemicals have attracted much attention due to their persistence, toxicity, bioaccumulation, and long-range transport potential. At present, the production and use of some of these toxic substances (for instance, OCPs) have been banned or restricted worldwide, and the demand for these compounds is expected to decline. However, production of the alternatives or replacements of these POPs, which are referred to as "Emerging Chemicals" with potentially toxic substances for humans, is projected to increase. Therefore, large amounts of emerging chemicals have been detected in municipal wastewaters, surface and ground waters, drinking waters as well as sediments and aquatic biota [4].

Emerging chemicals are also called emerging substances, emerging pollutants, or emerging contaminants. According to the definition given by the Network of reference laboratories, research centres and related organisations for monitoring of emerging environmental substances (NORMAN), emerging substances can be defined as chemical substances that have been detected in the environment, but which are currently not included in routine monitoring programmes. Moreover, the fate, behaviour, and (eco)toxicological effects of emerging substances are not well understood at present. Furthermore, emerging pollutants defined by NORMAN are pollutants that are currently not included in routine monitoring programmes, but may be candidates for future regulation. It depends on research data on their (eco) toxicity, potential health effects and public perception, and on monitoring data regarding their occurrence in the various environmental compartments. In short, emerging chemicals widely exist in the environment, potentially harmful to the ecological environment and human health, and also are the preferred substances for future environmental control. Examples of emerging chemicals include surfactants, flame retardants, pharmaceutical and personal care products (PPCPs), gasoline additives and their degradation products, biocides, polar pesticides and their degradation products, and various proven or suspected endocrine disrupting chemicals (EDCs).

Emerging environmental substances are not necessarily new chemicals. They are substances that have often long been present in the environment but whose presence and significance are only now being elucidated. NORMAN has identified a list of the currently and most frequently discussed emerging substances and emerging pollutants. Figure 1 shows NORMAN Prioritisation scheme for emerging substances. The currently and most frequently discussed emerging chemicals include phthalates, bisphenol A (BPA), perfluorinated compounds, polybrominated biphenyl ethers, hexabromocyclododecane, carbamate pesticide, etc. [5].

Emerging chemicals' effect on human health is a hot issue in the field of public health. Till now, new environmental pollution monitoring technologies are immature for most of the emerging chemicals. Some of them have no environmental monitoring methods (i.e., synthetic nano-and micro-plastics), and there is a huge gap in environmental exposure data. For example, US EPA conducted a statistical survey of 1400 chemicals at risk to health in the United States and found that only

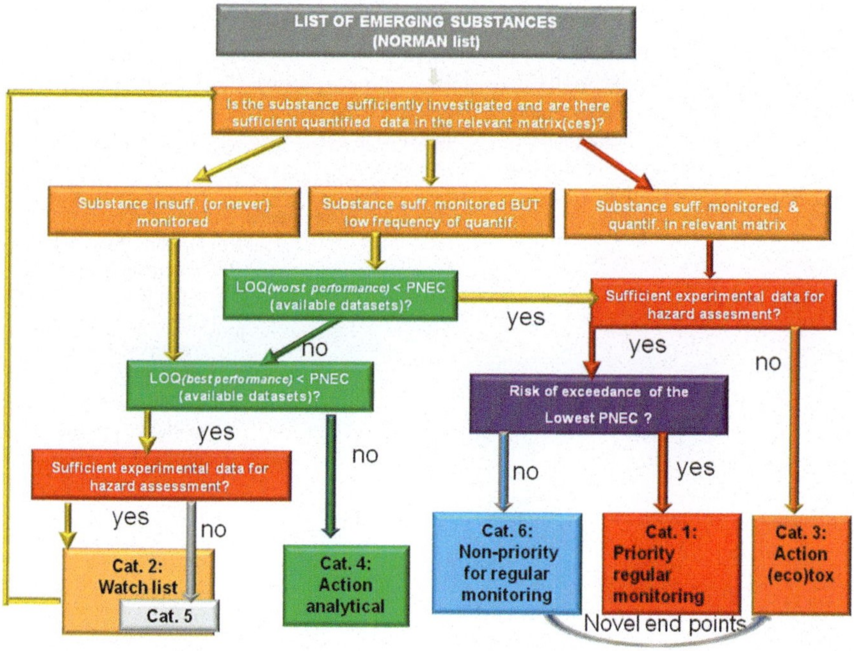

Fig. 1 NORMAN Prioritisation scheme for emerging substances [5]

6% of the population had exposure data, especially for sensitive populations (children, pregnant women, and the elderly). Meanwhile, the environmental and health hazard characteristics of most emerging chemicals are not clear, for instance, environmental targets and action modes of POPs and antibiotics. Table 1 shows the possible health and ecological effects of most frequently exposed emerging chemicals [6, 7].

Generally, declining sperm quality, infertility, higher rates of diabetes, cancer risks, and adiposidas are only some of the negative health effects associated with emerging chemicals. The effects of emerging chemicals are highly dependent on the time of exposure and absorption. The critical exposure window of emerging chemicals includes periods of pregnancy (mainly affecting fetal growth and development), infancy, early childhood and adolescence. During these critical phases, special protective barriers (blood–brain barrier, placenta barrier, blood–testis barrier, etc.) are not well developed to resist against many emerging chemicals in humans. Thus, even small doses of these chemicals can be detrimental to children development.

Currently, more than 200 emerging chemicals are known to have endocrine disrupting effects, such as bisphenol A, flame retardants, and phthalates. They are found in many products of everyday use, allowing them access into environment and human bodies with the usage of these products. Such emerging chemicals are persistent in environment and humans and can even have adverse effects at low concentrations. The consensus is that there is no safe dose for some emerging

Table 1 Health effects of the most frequently exposed emerging chemicals

Toxic chemicals	Possible health & environmental effects	Type	Application
BPA	Decline semen quality, impair fertility, disturb steroid hormones, and affect child development	Plasticiser, a main component in the manufacture of polycarbonate plastics, epoxies, and epoxy resins	Plastics, baby bottles and pacifiers, food packaging, such as internal coating of food cans, microwave dishes made of polycarbonate, plastic utensils, PVC saran wrap, computers, CDs, thermal paper (receipts), recycled paper towels and toilet paper, medical devices, dental composite fillings, adhesives, varnishes, nail polishes
Phthalates	Impair fertility, harmful for the unborn child	Softener for plastics	PVC flooring, plastic tiles, carpeting with plastic lining, vinyl wallpaper, shower curtains, leather furniture and bags, PVC food packaging, electrical cables, adhesives, paints, detergents, coated fabrics, medical products (gloves and hoses), cosmetics, textiles (water-resistant clothing, mud pants, plastic clogs), toys (like plastic dolls, water toys, kiddie pools, water wings, inflatable mattresses)
UV filters	Induce vitellogenin, alter the structure of reproductive organs, cause reproductive defects, disrupt gene expression, lead to individual malformation and death	Labelled as benzophenone-3, 3-benzylidene camphor, 4-methyl-benzylidene-camphor, 4,4-dihydoxy-benzopheone, benzophenone, ethylhexyl methoxycinnamate-parabens; in cosmetics: butylparaben, propylparaben; in foods: E214, E215, E218, E219	Body & baby care products

(continued)

Table 1 (continued)

Toxic chemicals	Possible health & environmental effects	Type	Application
Parabens	Lead to early puberty in females, fuel the growth of oestrogen-receptor positive breast cancers or impact fertility in males	in cosmetics: butylparaben, propylparaben; in foods: E214, E215, E218, E219	Body & baby care products, food, packaging, pharmaceuticals, tobacco, and shoe polishes
Flame retardants	Reduce peripheral thyroid hormone, increase the risk of obesity, injury neural behaviour		Children products, toys, textiles, clothes, footwear
Formaldehyde	Irritates mucous membranes and skin; can cause hypersensitivity, allergies, and cancer (nasal pathway)	Antibacterial, preservative	In many types of glues, lubricants and paints, plywood, particleboards furniture, laminate, textiles, and leather
Nonylphenol (ethoxylates)	Endocrine disruptors, persistent, accumulates in the environment	Tenside, surface active substance	Textile paper and pulp processing, paints, resins, glue, protective coating, detergents, animal medicine, pesticides, steal production
Heavy metals	Lead can cause brain damage, and cadmium can cause kidney damage. Both are classified as Carcinogenic, Mutagenic, or toxic for Reproduction (CMR)	Used in pigments in non-natural colourants	Paint and dye, PVC
Glycolethers	Classified as CMR can harm the unborn child, flammable, harmful when inhaled, in skin contact and swallowing	Solvents, detergents, pesticides, medicine, anti-freeze	Plaster paints, acrylic paints, enamel, wood varnish, glue, resins, detergents, paint and varnish remover, hydrolic liquids
Xylene and benzene (terpentine)	Harmful to reproduction, unborn child, causes skin irritation and eczema	Solvents, detergents, medicine, anti-freeze	Plaster paints, acrylic paints, enamel, wood varnish, glue, resins, detergents, paint and varnish remover, hydrolic liquids

(continued)

Table 1 (continued)

Toxic chemicals	Possible health & environmental effects	Type	Application
Polyvinyl chloride (PVC)	8-16 substances in PVC can cause different types of damage, such as dipentylphthalate (CMR, damaging to unborn child), cadmium, lead, and vinylchloride (carcinogenic, on combustion dioxin is released)	Synthetic material, soft or rigid, applied for many uses	Piping, textile (t-shirts) gloves, (building) plastics, toys
Organotins	Irritation of eyes and skin, toxic to reproductive system, on ingestion harmful to central nervous system, endocrine system and reproductive system	Works against bacterial growth, fungus and rot, and sweaty smell	PVC, furniture, t-shirt prints, diapers, tent cloth, textile, and leather
Brominated flame retardants	Persistent, bioaccumulative, and toxicant (PBT), some kinds also classified as CMR, some as endocrine-disrupting chemicals (EDCs)	Slows the combustion process by several minutes	Polystyrene, textile (furniture fabrics), stuffing of furniture and mattresses wrappings, isolation materials, plastics
Biocides	Toxic to living organisms skin irritation, burns, allergies	Intended to kill, deter, or render harmless organisms (plants and animals) considered harmful	Wood and wood products, fungicidal paint and the like

chemicals, including some heavy metals and POPs. For instance, lead is a kind of heavy metal with systemic multi-organ toxicity. Children aged 0–6 in rapid growth stage are at high risk of lead toxicity. Literatures show that a significant dose–effect relationship exists to the toxicity of lead to children. Higher exposure levels of lead (above 700 μg/L in blood) can cause serious damage to children's growth and development, and even lead to toxic encephalopathy or death. When the level of lead in children's blood is 100–200 μg/L, it is called subclinical lead exposure. It can also lead to hyperactivity, attention deficit, memory decline, cognitive, and behavioural impairment in children. Studies in the past 20 years show that lead exposure has no safety threshold for children's developmental toxicity. Even if the blood lead level of children under 6 years of age is less than 100 μg/L, it is still unsafe. 50 μg/L of the blood lead level in children is associated with the lower IQ, indicating that low lead exposure still damages the nervous system, cognitive, and behavioural development of children.

As the most vulnerable population, children are more sensitive to chemical substances because their entire body (immune system, nerve system, brain, organs) is developing and not yet mature and the barrier function is not completely formed. The baby's skin is not only more vulnerable but also more permeable than an adult skin. Children breathe much quicker, eat and drink much more than an adult. Therefore, children inhale and ingest more substances and residues per kilogram of bodyweight than an adult. More important, the body of children is not yet capable of detoxing all chemicals properly, keeping these chemicals or residues in their body longer. Slow metabolism prolongs the exposure of emerging chemicals in their bodies. Therefore, certain emerging chemicals that are considered 'safe' in adults can still be harmful for children.

Hazardous chemicals are found in toys, cosmetics, and other everyday products. So, babies and young children are exposed to numerous chemicals everyday through products they use. Prolonged contact – for example, while sleeping and playing on the floor – enhances penetration/inhalation into the body. Chemical substances can enter the body not only through the skin but also through inhalation and through mouth contact. The smallest amounts of these chemicals can cause long-term, sometimes lifelong, negative health effects. Studies show that some emerging chemicals are connected to childhood cancer and other chronic diseases. Moreover, the baby's development is primarily regulated through hormones, often in very small quantities. If a hormone-disrupting substance influences the process of development, there could be an irreversible impact in children's development. In this way, chemicals can change the signals of the endocrine system, and in the long run, have an effect on the reproductive organs, as well as on the behaviour of the child, which should be paid attention to in the future studies.

The ten chapters in this book focus on the most frequently exposed emerging chemicals in humans, and cover topics that range from occurrence and environmental risks through to the toxicological effects of eight kinds of emerging chemicals. Our aim in writing this book is to encourage further research in new directions by providing novel and provocative insights into the exposure assessment and potential mechanisms of emerging chemicals in humans. This book offers a state-of-the-art report on recent discoveries concerning emerging chemicals and where the field is going. It mainly focuses on advances made over the past 10 years on the exposure, metabolism, transformation, toxicity, molecular mechanism, and biomarkers of emerging chemicals in humans. Together, they have, to some extent, served to represent some of the important scientific questions and management issues that we need to consider in managing emerging chemicals. It will help graduate students and researchers to get an overall picture of emerging chemicals and may also devise methods to 'speed up' the risk assessment procedures and risk management of those compounds.

For protecting human health, we recommend some general tips avoiding toxic emerging chemicals at the end of this chapter, including:

1. Less is more: Think carefully about what you need, and use baby care products only if really necessary.

2. Always choose products very carefully: Pay attention to organic seals or labelling when purchasing products.
3. Avoid products with PVC: Food packaging made of glass, stainless steel, and porcelain is better than plastics.
4. At home: Pollutants can accumulate in indoor air, so air out regularly.
5. Eat fresh, and if possible, unpackaged organic foods: Do not use food from cans and do not heat food in plastic containers, especially not in microwave.
6. Pay attention to the ingredient information on cosmetic products: Leave care products that are conserved with propylparaben, butylparaben, or BPA, on the shelf, and avoid sun care products and skin care products with hormone-altering UV filters. Sun protection products with mineral UV filters are better.
7. Consider pacifiers, baby bottles, dishes, and other products that are BPA-free: Do not give your baby any soft plastic toys and do not buy toys with an unpleasant odour.
8. Choose preferably a room that is easy to air and ventilate (big windows, ventilation strip), is not situated too close to an open fireplace or the kitchen, is draft and mould-free, and start renovating activities well in time, preferably 3 months before moving into a new room.

References

1. Denison RA (2007) High hopes, low marks: a final report card on the high production volume chemical challenge. https://www.edf.org/health/reports/high-hopes-low-marks
2. Hartung T (2009) Toxicology for the twenty-first century. Nature 460(7252):208–212
3. Judson R, Richard A, Dix DJ et al (2009) The toxicity data landscape for environmental chemicals. Environ Health Perspect 117(5):685
4. PKS L, Yeung L (2017) Emerging chemicals of concern: preface. J Environ Sci 61:1–2
5. https://www.norman-network.net/sites/default/files/files/WG1/NORMAN_tree.jpg
6. http://www.projectnesting.org
7. http://www.wecf.eu

Exposure Assessment of Emerging Chemicals and Novel Screening Strategies

Qingyang Zhu and Haixia Dai

Abstract Persistent organic pollutants (POPs) commonly exist in various kinds of environmental mediums and can migrant into plant food sources and bioaccumulate in the fatty tissues of human body. Traditional approach to evaluate POPs in multiple biofluids is based on targeted analytic chemistry. Recently, the development of sophisticated analytical instruments (e.g., tandem mass spectrometry, MS–MS) has provided the opportunity to quantify and identify chemical compounds to achieve good sensitivity and selectivity. In this chapter, we discuss the current assessment tools of chemical pollutants, including classic targeted approaches and novel untargeted methods. Targeted biomonitoring studies typically focused on a specific group of interest chemicals such as phthalate, bisphenol A (BPA), and polybrominated diphenyl ethers (PBDEs). Recent studies tended to use noninvasive or less-invasive bio-matrices which could be accessible in sufficient amounts for the analysis and do not pose a health risk for the donor. There does not exist an ideal matrix for universal situations, but depending on the toxicokinetic of the targeted chemical. Exposome includes a series of quantitative and repeated metrics of both endogenous and exogenous exposures that describe, holistically, environmental influences or exposure over a lifetime. At the current stage, the exposome is still in its infancy. Many technical and statistical challenges remain unsolved. Combined with data mining, via a series of statistical approaches, exposome shows great potential in identifying markers that can further lead to targeted analyses.

Keywords Persistent organic pollutants · Exposure assessment · Phthalates · Bisphenol A · Polybrominated diphenyl ethers · Exposome

The original version of this chapter was revised: Acknowledgment has been included. The correction to this chapter is available at https://doi.org/10.1007/978-981-32-9535-3_11

Q. Zhu
School of Public Health, Fudan University, Shanghai, China

Key Laboratory of Public Health Safety, Ministry of Education, Fudan University, Shanghai, China

H. Dai (✉)
State Environmental Protection Key Laboratory of Formation and Prevention of Urban Air Pollution Complex, Shanghai Academy of Environmental Sciences, Shanghai, China
e-mail: daihx@saes.sh.cn

Y. Zhang (ed.), *Emerging Chemicals and Human Health*,
https://doi.org/10.1007/978-981-32-9535-3_2

1 Introduction

Persistent organic pollutants (POPs) commonly exist in various kinds of environmental mediums. Due to their high resistance to degradation and high persistence in the environment, POPs can migrate into plant food sources from the environment and then bioaccumulate in the fatty tissues of human body with potential adverse health effect [1].

The traditional and the most widely used approach to evaluate POPs in multiple biofluids is based on targeted analytic chemistry. Recently, the development of sophisticated analytical instruments (e.g., tandem mass spectrometry, MS–MS) has provided the opportunity to quantify and identify chemical compounds to achieve good sensitivity and selectivity. However, targeted approach inevitably ignores the presence of other POPs contaminants since it requires standard references and analyze specific information. As a result, it is not able to give an exhaustive overview of other organic components in the biofluids. Full-spectrum acquisition techniques that provide accurate high resolution mass spectrometry, such as gas chromatography-mass spectrometry (GC-MS) and liquid chromatography-mass spectrometry (LC-MS), show great potential for obtaining information on large amounts of organic compounds in human body.

In 2005, Wild proposed the conception of exposome, representing the totality of exposures throughout the lifespan [2]. Research efforts have soon begun after the definition was originally made, leading to a revised working definition that the exposome represents the cumulative measure of exposures to both chemical and nonchemical agents such as diet, stress, and socio-demographic factors. Exposome includes a series of quantitative and repeated metrics of both endogenous and exogenous exposures that describe, holistically, environmental influences or exposure over a lifetime. It means that not only the traditional measures of exposure (e.g., traditional biomonitoring, environmental monitoring) but also untargeted discovery of unknown chemicals of biological importance are included [3, 4]. Going a step beyond traditional biomonitoring, exposomic approaches aim to capture all exposures that potentially affect health and disease. Biomonitoring serves as the key tool to define exposure disease risks given the biological significance of internal exposure measurements. With the development of modern technology, novel screening strategies are critical in achieving a comprehensive understanding of exposures that are vital in personal and public health. Full understandings of the complex interactions between genetics and environmental exposures may greatly help to solve the mysteries of the epidemiology, trends, and prevention of many diseases.

We are only beginners to understand the complexities of environmental exposures and their impacts on human health. At present, we have limited estimates of the impact of environmental exposures on health, and uncertainty greatly exists in those [5]. In this chapter, we are going to discuss the current assessment tools of chemical pollutants, including classic targeted approaches and novel un-targeted methods. In addition, the current progress of exposome studies as well as its screening strategies are described here.

2 Targeted Approaches

Traditional biological measurements, also called targeted analyses, measure a target chemical, metabolite or reaction product in biofluids such as urine or blood. In epidemiological studies, targeted approaches have become a key component of exposure assessment to characterize biologically persistent chemicals, whereas short-lived chemicals are effectively measured only if the individual is undergoing continuous exposure, if the exposure timing is known. The utility of targeted methods in characterizing phthalates, bisphenol A, and PBDEs are discussed in this section.

1. **Phthalates**

Phthalates, the diesters of 1,2-benzendedicarboxylic acid (phthalic acid), are a class of synthetic chemicals that are used widely in industrial applications. Human exposure is common due to the widespread use of phthalates in commodities [6]. Ingestion, inhalation, and dermal contact are considered important routes of phthalates exposure for the general population, but the sources, pathways and routes may vary by phthalates due to the differences in their utility in products or their chemical and physical properties [7]. Phthalates have short biologic half-lives and are quickly excreted from the body.

Previous studies mostly focus on seven specific kinds of phthalates: (1) diethyl phthalate (DEP); (2) di-n-butyl phthalate (DBP); (3) di-isobutyl phthalate (DiBP); (4) butyl benzyl phthalate (BBzP); (5) di(2-ethylhexyl) phthalate (DEHP); (6) diisononyl phthalate (DiNP); and (7) di-isodecyl phthalate (DiDP). Their chemical structures are shown in Fig. 1. There are both commonalities and differences among

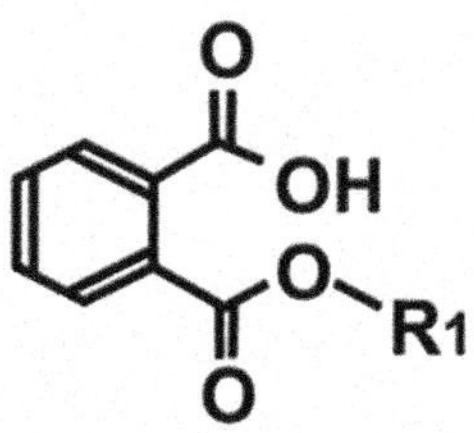

Monomethyl phthalate [R_1: CH_3]
Monoethyl phthalate [R_1: CH_2CH_3]
Monoisobutyl phthalate [R_1: $CH_2CH(CH_3)_2$]
Mono-*n*-butyl phthalate [R_1: $(CH_2)_3CH_3$]
Mono-*n*-pentyl phthalate [R_1: $(CH_2)_4CH_3$]
Mono-*n*-hexyl phthalate [R_1: $(CH_2)_5CH_3$]
Monocyclohexyl phthalate [R_1: C_6H_{11}]
Monobenzyl phthalate [R_1: C_7H_7]
Mono(2-ethyl-5-carboxypentyl) phthalate [R_1:$CH_2CH(C_2H_5)C_3H_6COOH$]
Monoisopropyl phthalate [R_1: $CH(CH_3)_2$] (internal standard)

Fig. 1 Chemical structures of urinary phthalate metabolites [9]

phthalates in their uses and thus sources of exposure. Following exposure and uptake, phthalates are rapidly metabolized and excreted in urine and feces. They typically undergo phase I hydrolysis followed by phase II conjugation in human body [8]. In phase I, the phthalate diesters are hydrolyzed into monoester metabolites by lipases and esterases in the intestinal epithelium, live, blood, or other tissues. The monoester metabolites then undergo phase II biotransformation catalyzed by uridine 5′-diphosphate glucuronosyltransferases (UGTs) to form glucuronide-conjugated monoesters that are excreted in urine. Thus, detecting urinary biomarkers has been the most common method for assessing phthalate exposure in environmental epidemiology studies to date.

Exposure biomarkers have the advantage of being a measure of internal dose, which could account for all routes of exposure. This is desirable in epidemiology studies of chemicals with multiple exposure pathways. Many different types of biospecimens have been used to assess the environmental chemical exposure. Phthalates have been measured in urine and blood (serum, plasma or whole blood) as well as saliva, semen, breast milk, swear, amniotic fluid, and umbilical cord blood. However, measured concentrations in urine are normally much higher than the concentrations in other specimens. Measuring phthalates in urine offers many advantages over measuring the diesters or their metabolites in blood, which include (1) ease of sample collection; (2) larger sample volume; (3) higher concentrations of the metabolites; and (4) reduced potential for contamination by the parent diester and subsequent formation of metabolites by enzymes [10]. Moreover, for some phthalates, such as DEHP, the monoester metabolites are thought to be more biologically active than the parent diesters.

Moderate to strong correlations between urine and serum concentrations have been reported for the secondary metabolites of DEHP and for the primary metabolite of DEP [11, 12]. Oral dosing studies in human have shown that DEHP concentrations peak in serum within approximately 2 h of exposure and decline rapidly [13]. Reflexively, urinary metabolite concentrations rise, with almost all of orally administered dose excreted within 24 h after exposure. At the individual level, significant correlations between phthalate metabolite concentrations in serum and urine may not be expected, but moderate to strong correlations in these two matrices are plausible at the population level.

Phthalate-related epidemiological studies normally examine a large number of metabolites due to the multi-analyte nature of the most widely used assays in measuring the metabolites of phthalate. Various approaches have been proposed to reduce the number of exposure variables in these diseases. These approaches include (1) summing phthalate metabolites of "low molecular weight" and "high molecular weight"; (2) summing metabolites of the same parent compound; and (3) summing all measured metabolites regardless of parent chemicals. Though the application of these approaches could successfully reduce the number of statistical comparisons involved, they also have some obvious disadvantages. The approaches above rely greatly on an assumption that the chemicals or metabolites being summed have similar biological activities and targets. Nevertheless, this may not be plausible, even if the metabolites are summarized from the same parent phthalate while the monoester itself is the most biologically active form. Similarly, when more nuanced approaches are applied to weight the toxicities of single metabolites based on combinations of the empirical data, there is an assumption that the same basis being used for the weightings represents the most sensitive endpoint for each individual phthalate.

Measuring phthalate metabolites in urine is currently the most common method of assessing exposure in epidemiological studies. The advantage of this approach is the wide availability of sensitive analytical methods, where most phthalate metabolites represent the specificity of a particular parent chemical, and the long biological half-life of monoester metabolites compared to the respective phthalate diester typically measured in serum. The possibility of contamination of enzymatic activity is reduced compared to the measurement of parent chemical or metabolites in other biological fluids. Limitations of this approach include consideration of half-lives on the order or hour, consideration of urine dilution (and uncertainty of the most appropriate method for urine dilution), and metabolite concentrations measured in urine represent only an approximation for the dose at the target site. Moreover, individual biomarker measurements do not provide information on the exposure pathways required for risk management. Studies that assess sensitivity and specificity have shown that a single measurement may possibly be able to reasonably predict long-term exposure categories (i.e., months, with limited data for years). However, collecting additional samples will improve sensitivity, and researchers should strive to conduct studies that collect multiple samples for measuring phthalate metabolites in the exposure window that is most relevant to the results of interest.

In summary, the measurement of phthalate metabolite concentrations in urine can be used as an effective method for estimating exposure in epidemiological studies. However, the advantages and limitations of this approach need to be carefully considered when interpreting the results of these studies.

2. Bisphenol A

Bisphenol A (BPA), a common industrial chemical synthesized by condensation of two phenol groups and one acetone molecule, is often used in food and beverage packaging, medical equipment, electronics, flame retardants, adhesive, building materials, automobiles, and paper coatings [14]. The first commercial production of BPA was in the United States in 1975. In the following 60 years, BPA becomes ubiquitous in the environment, thus increasing the likelihood for human exposure.

Since a large number of studies have reported the estrogenic properties of BPA [15], it is described as an endocrine disruptor chemical (EDC). BPA is capable of binding and activating human estrogen receptor, but with a capacity of 1000–5000 times less than the endogenous 17-β-estradiol. BPA has been shown to interact with other endocrine receptors including thyroid hormone receptors and peroxisome proliferator-activated receptor gamma [16]. BPA is classified as a reproductive toxic substance of category 3 as an alarming substance for human fertility.

Biomonitoring of BPA in various human matrices has been reviewed previously [17]. Urine is traditionally the most preferred matrix because BPA is extensively bound by glucuronidation and excreted in the urine, and sampling is minimally

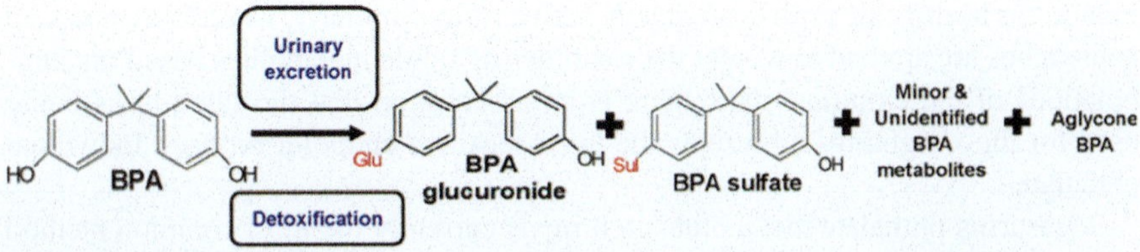

Fig. 2 BPA and its conjugates [20]

invasive. However, given the short half-life of BPA in human body(~6 h), the observed levels in urine could only reflect recent exposure, limiting its utility as biomarker [18]. Current analytical methods determine the total BPA concentrations after enzymatic hydrolysis preferably at 37 °C for a few hours and overnight in some specific situations. Typically, only β-glucuronidase is used in the de-conjugation step since BPA is predominantly present in the form of a glucuronide conjugate, while few have additionally used sulfatase enzyme for the release of BPA from the sulfate conjugate that occurs as a very minor fraction. In addition to the glucuronide form, the sulfatase will also be uncoupled if the β-glucuronidase is derived from *Helix pomatia*-H1 compared to *Escherichiacoli*-K12 [19]. Some of the problems leading to inaccurate or underestimated measurements may be due to suboptimal conditions of total BPA that arose from insufficient enzyme concentrations, improper selection of enzymes, incomplete deconjugation, unfavorable hydrolysis conditions and overall suboptimal de-conjugation. The structures of BPA and its conjugates are illustrated in Fig. 2.

Glucuronide and sulfate conjugates are the most common metabolites of xenobiotics and exogenous chemicals that undergo phase II metabolism in humans as a detoxification measure. Such conjugates are preferably detected and measured using LC-MS based methods. Several studies have reported the simultaneous determination of aglycone BPA and its major conjugates BPAG, BPAs and BPADS. Until recently, research was limited by the lack of custom synthesis of BPA conjugates and the commercialization of the respective labeled internal standards. Although some studies have reported the analysis of conjugated BPA in bio-specimens, the information was derived from taking the difference between total and free BPA rather than directly measuring BPAG and BPAS. Previous studies used urine [21], plasma [22], serum, and cord serum [23] to measure BPA and its conjugates. Extraction and satisfactory recovery of BPA and its conjugates from biological samples is critical, followed by their optimal separation from matrix ions, which produces optimal ionization, detection, and quantification through the MS platform.

Most studies used liquid chromatography to separate aglycone BPA and its conjugates. All reported methods used a gradient elution of concentrated and unaltered solvents made with acetonitrile or methanol [22]. The run length and retention times of aglycone BPA and its conjugates depend on the chromatographic conditions used in these studies. Run time ranged between <5.0 min and 22.0 min. The BPA conjugate elutes before the aglycone BPA and therefore has a shorter retention time compared to the latter. Electrospray ionization (ESI) mode was the only optimized ionization mode. The presence of a phenolic hydroxyl group in the aglycone BPA and its conjugate promotes efficient ionization in the negative electrospray ionization mode and is therefore preferred. In the ESI negative mode, most studies used the following multiple reaction monitoring mass conversions to accurately detect and quantify BPA: m/z 227–2012; BPAG: m/z 403–113; BPAS: m/z 307–227; and BPADS: m/z 387–307. Isotope-labeled internal standards are used in the initial stages of sample preparation to ensure that quantification is not significantly affected by matrix effects.

Total BPA measurement in urine is a reliable measure only when external contamination is controlled. Since the introduction of BPA from an external source not only increases the aglycone but also increases the total BPA level, measuring either one is not necessarily a reliable biomarker of exposure to BPA. However, BPAG and BPAS quantification are gaining consideration and value because the levels are not affected by the general sample processing procedures either during collection or analyses. BPA and its conjugates in human matrices are primarily quantified as an integral part of method development, and very few studies have applied this method to biological monitoring of human exposure to BPA. However, BPAG and BPAS quantification are gaining consideration and value because the levels are also not significantly affected by the general sample processing procedures during collection. BPA and its conjugates in human matrices are primarily quantified as an integral part of method development, and very few studies have applied this approach for biomonitoring of human exposures to BPA. The specificity of BPAG is similar to aglycone BPA in exposure assessment studies [24–26]; however, the frequency of detection of BPAG in urine is almost always greater than aglycone BPA. Among the populations studied, BPAG was measured in almost all human matrices and was measured at higher detection frequencies and concentrations than aglycone BPA.

Urine BPAG concentrations account for 90–95% of total BPA, while aglycone BPA accounts for only 1–2% [27]. In contrast, one study reported that BPAG and aglycone BPA accounted for 57% and 32% of total BPA in urine [21]. The lower proportion of BPAG in this study was attributed to the possible external contamination of BA samples and the underestimation of BPAG using inappropriate internal standards (labeled BPA instead of BPAG). In the case of umbilical cord serum, BPAS is frequently detected and is a major fraction of the total BPA of 45%, followed by 36% aglycone BPA and 19% BPAG [28]. These conversion factors are very important when directly measuring BPAG and BPAS before comparing BPAG and BPAS with BPA exposure data.

The main limitation of using BPAG and BPAS as additional biomarkers for exposure to BPA is that if the urine sample is stored at room temperature, the conjugated form can degrade within a day or two [29]. This leads to underestimation of BPA exposure when the individual conjugates are analyzed and measured directly. BPAG is stable at −80 ° C for a total duration of 28 days and is stable in at least three freeze and thaw cycles [30]. While urine samples are properly stored, they may be exposed to ambient temperatures for a period of time during collection, handling, and transportation, which may potentially compromise the utility of these surrogate biomarkers. When total BPA is used as a biomarker, filled blanks and reagent blanks are used as quality control methods to control background contamination of BPA, but no such measurements can be used to estimate degradation of urine samples between sample collection and pre-analysis. Other potential limitations of the method include: (1) lack of documented statistically significant, positive, and linear correlations between BPAG and total BPA; and (2) unknown and unrecognized outside BPAG and BPAS BPA conjugates are unmonitored and unmeasured.

Although BPA conjugate measurements are possible due to recent advances in analytical methods, there are some limitations and challenges that need to be overcome before considering their application in human biomonitoring research. If successful, analysis of BPA conjugates as additional biomarkers may help to better understand metabolic and distributional differences in life stages, inter-individual differences, and potential health effects of biologically active forms.

3. PBDEs

Since the 1970s, polybrominated diphenyl ethers (PBDEs) have been produced due to the ban on the use of previously used flame retardants. They are used as additives in a variety of polymers and are found in a variety of consumer products, including furniture, electronics, and automotive parts. PBDEs are environmental contaminants with persistent and bioaccumulative properties. Studies have detected PBDEs in human tissue samples such as placenta, adipose tissue, breast milk, and blood. The molecular structure of PBDEs is shown in Fig. 3.

Human exposure to PBDEs is thought to result from dietary sources, including fish, fatty foods, and breast milk. However, oral ingestion from dust and leachates may be a larger source, particularly for children. Once absorbed, PBDEs rapidly distribute into body fat. The metabolism and elimination of PBDEs in humans has not been well characterized. Some studies have shown that the elimination half-life of decabromodiphenyl ether is 11–18 days and the half-life of octabromodiphenyl ether homologs is 37–91 days [31]. In animals, PBDE elimination occurs primarily through fecal excretion, with decabromodiphenyl ether being eliminated more rapidly than other less brominated PBDEs.

Polybrominated diphenyl ether homologs for biological monitoring typically include three bromines (BDE-17, BDE-28), four bromines (BDE-47, BDE-66), five bromines (BDE-85, BDE-99, BDE-100), hexabromo (BDE-153, BDE-154), and heptabromo (BDE-183). Due to its low lipid content, low organic contaminant

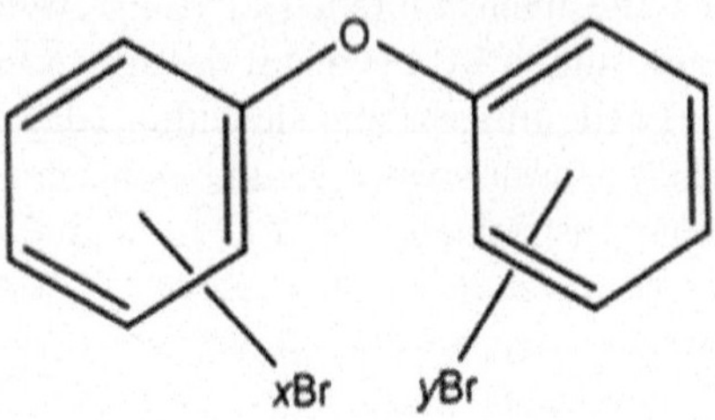

Fig. 3 The molecular structure of PBDEs ($x + y = 1\sim10$)

concentrations (usually below 1 ng/g), aqueous properties and high concentrations of hydrophobic proteins, many laboratories have demonstrated that blood analysis is difficult. Improved extraction efficiency, cleaning methods, and instrument sensitivity are needed to solve these problems.

Historically, liquid–liquid extraction (LLE) methods have been used most frequently to extract organohalogen contaminants like PBDEs from blood [32]. LLE allows for the extraction of large volumes of different matrices (whole blood, plasma, or serum) at relatively low extraction costs. However, this method is labor-intensive and does not allow for high throughput analysis. Solid-phase extraction has recently become a common method of blood extraction, and automated SPE has proven to be useful for large-scale monitoring projects using human blood samples. Microwave-assisted extraction is an alternative to LLE and SPE, but organic pollutants from blood have never been explored before. Cavity-dispersed microwave-assisted extraction (MAE) is widely used to analyze heavy metals in biological samples and to extract organochlorine contaminants from solid biological tissues and should be suitable for extracting PBDEs from blood matrices. In MAE, multiple samples are placed in an oven where microwave energy is dispersed throughout the oven cavity, while a focused microwave (FME) system applies all of the microwave energy to a single sample in a smaller over cavity, which leads to higher overall energy transfer to the sample compared to MAE. Among all the above methods, FME was found to be the most reliable one for extracting PBDEs from serum. It may result in the highest IS recovery and low to moderate variability without affecting the measured PBDE concentration. All methods except FME have quantitative problems with PBDEs. Furthermore, the highly reproducible concentrated microwave energy from this method may be the reason for its optimum performance. In addition, automation of extraction procedures, reduced solvent requirements and convenience, and reduced labor in the use of disposable glass extraction vessels may also be logistical reasons [33].

Measuring PBDEs in breast milk is vital to study the adverse health effect of early life exposure. PBDEs are lipophilic and can easily accumulate in breast milk. Hence, a major source of exposure to PBDEs for breastfed infants in early postnatal life is through the breast milk. Lin et al. introduced a tandem SPE cleanup method based on sulfuric acid silica cartridge coupled with basic alumina cartridge to measure PBDEs in human milk sample [34]. In contrast to the automated sample clean-up system (ASCS), SPE method presented comparable specificity with highly reduced consumption of solvents which results in much lower background in the procedural blank, reduced time, and enhanced sample pretreatment throughput. However, no present study has compared the performance of SPE method and FME method in extracting PBDEs from human breast milk, which warrants further investigation.

In addition, human hair and nails are good examples of noninvasive matrices. These matrices are easy and inexpensive to access and collect compared to other biological samples and are easy to transport and store. In addition, these matrices are stable with low or no active metabolic activities. Once the chemicals are incorporated into the keratinous tissue of hair or nails, the levels remain relatively uncharged because both of them are slowly growing tissues. Therefore, they provide the opportunity to measure exposures in a time window from 1 week to 1 month, which could reflect past and present exposures. The correlations of the concentrations of PBDEs in serum with those in hair/nails were tested previously, and the results showed that significant positive correlations were observed for the levels of individual BDE congeners (e.g., BDE-28, -47, -99, -100, and -154). However, no significant associations between the levels of BDE-209, TBB, or TBPH in hair, fingernail, and toenail versus serum were observed. Rapid metabolism and elimination are a plausible reason. Due to the metabolism in the body, it is unlikely that BDE-209 or TBPH can be transported through the body and reach hair and nails. The high levels of them in hair and nail samples may come from external sources such as absorption from air and dust [35].

Though serum PBDEs levels could better reflect human exposure, the use of noninvasive matrices, such as on hair and nails, is highlighted due to the easy accessibility and collectability. Moreover, PBDEs in breast milk is also an important indicator to analyze the adverse effect of early life exposure. The quantification of PBDE species with short half-life like BDE-2019 still needs to be further investigated.

3 Nontargeted Approaches and Exposome

Only a few hundred chemicals are routinely assessable through targeted methods. The exposomic approaches are critical to understanding the thousands of chemicals which people are exposed to everyday through direct chemical exposure consequences (e.g., cortisol levels due to exposure stress or noise). Through nontargeted biomonitoring approaches like high-resolution metabolomics, thousands of compounds can be monitored with a relatively small amount of biological specimen (about 100 μL) and for the cost of a single traditional biomonitoring analysis of around 10 target chemicals [5].

Nontargeted analyses of small molecules or macromolecular adducts could be conducted in various kinds of biospecimens. Such methods compare features of thousands of chemical features which are similar to ions with a given mass-to-charge ratio and a specified retention time in traditional biomonitoring process. Blood, urine, and other matrices are well suited for exposome-wide association studies (EWAS). Indeed, untargeted analyses performed using the current generation of liquid chromatography-high resolution mass spectrometers can detect over 30,000 small-molecule features [36] and over 100 human serum albumin (HSA) adducts of reactive electrophilic chemicals at the nucleophilic locus Cys34 [37]. Dealing with the rich datasets derived from untargeted analyses of multiple biospecimens offers a solution for discovering harmful exposures that have escaped scrutiny so far. It is also vital to know that full annotation of molecular features is a necessary part for case–control comparisons while LC-HRMS signatures are available. Archived biospecimens from well-designed cohort studies already exist. With continued advancement in untargeted analyses, there is potential to make significant advances in human health through uncovering unknown exposures [38, 39].

However, challenges exist with detecting chemicals, especially for those at low levels. It is also hard to define reference doses of "normal" exposure and linking these measures to active intervention approaches. Because blood xenobiotics concentrations tend to be greatly lower than the chemicals derived from drugs, food, or endogenous sources, nontargeted analyses of xenobiotics are not as efficient and reliable as those of ingested and endogenous sources for detecting many exposures of interest [40]. To determine the health impacts of these exposures, it is necessary to develop semi-targeted or multiplexed methods that increase the signals of exogenous molecules relative to those of endogenous origin. Analyzing suspected chemicals of concern (or suspect screening) can be prioritized through measuring groups of chemicals with known biological effects, but specific assumptions about toxicological pathways cannot be determined. Exposomic biomonitoring, including untargeted and suspected chemical analyses, has offered extraordinary potential for increased understanding of complicated chemical exposures.

When using the exposome approaches, special attention must be paid to which matrices can be practically collected and which matrices are associated with the assessment of chemical. In general, the least invasive specimens, such as blood and urine, are the preferred matrix in which the chemicals could be conveniently collected [41]. Although most exposure analyses are performed with urine or blood samples as they can be easily collected, there are other sample types which are also valuable in exposome interrogation. For example, saliva can be collected from school-age children and adults but is also a problematic matrix to collect from newborns because of choking dangers associated with the collection devices. Even if the matrix can be easily collected without invasion, it may not contain the target chemical or potential chemicals for many reasons, such as protein binding that prevent their secretion [42]. In addition, in nonsterile saliva, the contribution of the oral microbiome can affect the composition of the analytes to be tested. Buccal and nasal swabs are also used to assess the biological consequences of external exposure. In those samples, DNA, mRNA, and their adducts have been the main focus to date [43], but these sample types could also be significantly affected by the presence of powerful microbial communities that can affect the composition of exposed components.

When selecting samples for exposome-type research, an important consideration is the expected presence of a particular chemical in the harvested sample. Because chemicals may exhibit unexpected pharmacodynamics and biotransformation, it could be essential to collect multiple sample types from each individual to define the exposome. Blood circulates throughout the body, so its assessment has an advantage because it has been exposed to countless ways in which environmental chemicals enter the body. However, it is known that some analytes could specifically accumulate in specific tissues. Therefore, a broad-spectrum assessment of samples collected from multiple patient samples will provide the best insight into exposome.

The identification of unknown chemicals remains a major challenge for understanding the exposome. Research should try to prioritize the development of methods for determining relevant exposures and identify the sources of specific chemical characteristics. By correlating changes in microbiome, metabolomes, proteomics, etc. with unknown analytes, we can begin to determine the profile of the unknown toxic exposure and its consequences. In addition, biomonitoring techniques that can assess changes in cell composition or cell developmental capacity may identify the risk of subsequent health conditions such as cancer and neurodegenerative diseases. Linking unknown exposure to potential disease outcomes can further support understanding the resource inputs required to accumulate lifetime exposure.

Exposome analyses normally based on high-throughput advanced mass spectrometry technology. Analyzing bio-matrices to acquire the totality of exposure to organic pollutants is a small part of the assessment of the large and complex internal chemical environment consisting of exogenous reactions and a component for expanding human exposome [44].

Advanced high-resolution mass spectrometers (MS) allow expanded metabolic detection, leading to the exploration of a wider, untargeted chemical space in the exposome [45]. Andra et al. proposed a workflow for the analysis using the analysis of small molecule and metabolites as an example. This workflow including identification, quantification, and integration of the exposome into health studies (Fig. 4). The generic approach has two consecutive aspects: First, using nontargeted or "discovery" methods that employ high resolution mass spectrometry detection following liquid chromatography (LC) or gas chromatography (GC) separation to generate one large mass spectrometry data set; second, large-scale fragmentation of organic compounds and biomolecules which could be affected by environmental exposures. This approach could be combined with data mining via a series of statistical approaches and are useful in identifying markers that can then lead to targeted analytes.

The annotation of unknown chemical species could be extremely time-consuming and therefore only done on specific features. Through the joint efforts of many research groups to identify, catalog and disseminate information related to newly discovered small molecules, the limitations on chemical annotations can best be overcome. In addition, continuing to focus on bioinformatics techniques to extract information about important chemical characteristics will allow for semi-targeted methods for unknown and low abundance chemicals.

The omics technique has the potential to discover unknown compounds. Through the continuous advancement of mass spectrometry, low abundance chemicals can be located and characterized. By comprehensive coverage of the metabolome, the reference metabolic profile combined with health outcome data will provide a baseline for identifying unknown analytes with health relevance. Identifying and cataloging unknown analytes through cross-laboratory efforts will be a real task in advancing the exposome.

The application of nontargeted metabolomics to identify environmental exposures associated with human health is facing some unique challenges. The largest reference databased for metabolomics including the Metabolite and Tandem MS Database (METLIN) and the Human Metabolome Database (HMDB). Both databases focus on naturally occurring metabolites. The number of compounds in METLIN and HMDB that may be relevant to exposure studies has not been well estimated yet. The number of databases available for metabolomics is continuously expanding and has unique utility based on research questions. A broader discussion of metabolomics database resources can be made. To facilitate large-scale exposure

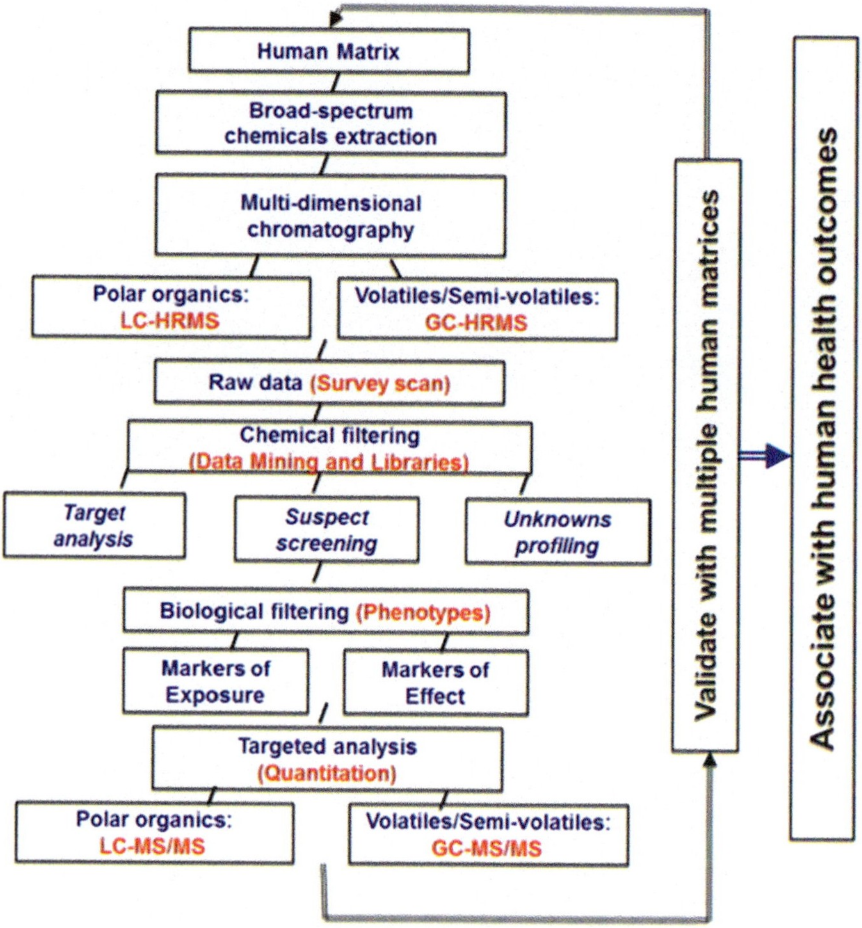

Fig. 4 An overview of applying high resolution mass spectrometry in human biomonitoring studies [46]

omics research, the field can benefit from owning a database or having a database search function dedicated to environmentally exposed chemicals.

Manrai et al. described that the human exposome data structure is a high-dimensional set of highly heterogeneous exposure variables that may change during repeated sampling in an individual's lifetime (Fig. 5a) [47]. Time-dependent and high-throughput genomic-scale phenotypic data types such as gene expression, protein expression, and metabolomics data are structurally similar.

However, there still a gap in assaying to operationalize an exposome-wide measurement platform. There are still signifijcant challenges in the field of metabolomics, such as determining the exact chemical identity or structure of an analyte in human tissue detected from mass spectrometry. In addition, another important challenge is the data integration across external and internal exposed domains. For

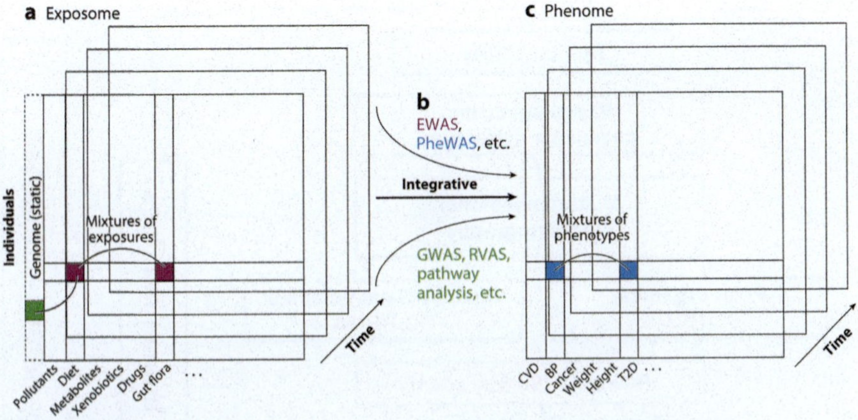

AR Manrai AK, et al. 2017.
Annu. Rev. Public Health. 38:279–94

Fig. 5 Data analyses in high-throughput exposomic studies [47]

example, how does sensor and physical measurements or socio-demographic factors at the individual level affect the results? How the distribution of internal exposure indicators affects the distribution of internal exposure indicators? Data integration across these domains is typically achieved by merging across space–time coordinates, a resource-intensive but often straightforward database task. Integrating the multidimensional and vertical data streams that appear in external and internal exposure measurements remains a huge challenge. Methods such as Specification Correlation Analysis or Graphical Minimum Angle Regression (Graphical LASSO) [48] techniques can enable investigators to map on large dataset to another, but methods are missing to consider longitudinal data. Computational methods enjoying a resurgence in the data science community, such as neural networks, may also be harnessed to assimilate data over different dimensions [49].

Information to support the interpretation of primary findings will require the continued development of ontologies and databases that are relevant to the exposome. Two established databases will help to develop standards for naming and toxicology knowledge, namely the Toxic Exposome Database (T3DB, http://www.t3db.ca) and the Comparative Toxico-genomics Database (CTD, http://ctdbase.org). T3DB currently provides information on more than 3600 compounds and over 2000 targets, expression datasets of more than 15,000 genes, and extensive information on chemical concentrations in biological fluids and reference chemical spectroscopy data. Software libraries such as GitHub (http://github.com), development tools such as WebProtege, and portals such as Bioportal can implement public and open source development of ontology. Discovery research in contact with humans is a big data analytics integration challenge involving statistics, computer science, biomedical, and public health. Further investigations need to make full use of the information in the high-dimensional exposure data.

4 Summary

In this chapter, we briefly discussed the current status of the application of both targeted approaches and nontargeted exposomic methods in the exposure assessment of chemical pollutants. Traditional biomonitoring studies typically focused on a specific group of interest chemicals such as phthalates, BPAs, and PBDEs. Recent studies tended to use noninvasive or less-invasive bio-matrices which could be accessible in sufficient amounts for the analysis and do not pose a health risk for the donor. There is not an ideal matrix for universal situations, but depending on the target chemical, toxicokinetic of the chemical, LODs, available amounts, and so on. After exposure, chemicals leave traces in human organism. For some compounds, this trace can be easily detected but not for others. Therefore, new exposure or effect biomarkers must be continuously researched and improved to detect this kind of traces. The development of group biomarkers is particularly important for targeted biomonitoring studies which are designed for those purposes.

At the current stage, targeted approaches are indispensable due to the huge implemental and statistical challenges that nontargeted exposomic studies are facing. The exposome concept is both a challenge and a potential driver for environmental health studies. The exposome is still in its infancy. Many technical and statistical challenges remain unsolved. In the case of assessment, hundreds of time-varying exposures need to be accurately assessed. Nevertheless, increasing the number of exposures assessed should not be done at the cost of increased misclassification. Estimating the association between a large amount of exposures and health could lead to statistical problems, including the increasing false discovery rate as well as the difficulty in efficiently untangling the exposure truly affecting the health outcome as they are significantly correlated with each other. These challenges should not hinder research that relies on the exposome methods, and the ongoing projects will hopefully allow advancement.

Acknowledgment This work was modified from the paper published in *Environmental Health Perspectives* [50] and *Environment International* [51]. The related contents are reused with permission.

References

1. Jones KC, De Voogt P (1999) Persistent organic pollutants (pops): state of the science[J]. Environ Pollut 100(1–3):209–221
2. Wild CP (2005) Complementing the genome with an "exposome": the outstanding challenge of environmental exposure measurement in molecular epidemiology[J]. Cancer Epidemiol Biomark Prev 14(8):1847–1850
3. Wild CP (2012) The exposome: from concept to utility[J]. Int J Epidemiol 41(1):24–32
4. Miller GW, Jones DP (2014) The nature of nurture: refining the definition of the exposome[J]. Toxicol Sci 137(1):1
5. Jones DP (2016) Sequencing the exposome: a call to action[J]. Toxicol Rep 3:29–45

6. Crinnion WJ (2010) The cdc fourth national report on human exposure to environmental chemicals: what it tells us about our toxic burden and how it assists environmental medicine physicians[J]. Altern Med Rev 15(2):101–108
7. Meeker JD, Sathyanarayana S, Swan SH (2009) Phthalates and other additives in plastics: human exposure and associated health outcomes[J]. Philos Trans R Soc B Biol Sci 364(1526):2097–2113
8. Frederiksen H, Skakkebaek NE, Andersson AM (2007) Metabolism of phthalates in humans[J]. Mol Nutr Food Res 51(7):899–911
9. Yoshida T (2017) Analytical method for urinary metabolites as biomarkers for monitoring exposure to phthalates by gas chromatography/mass spectrometry[J]. *Biomed Chromatogr* 31(7):e3910
10. Koch HM, Calafat AM (2009) Human body burdens of chemicals used in plastic manufacture[J]. Philos Trans R Soc B Biol Sci 364(1526):2063–2078
11. Frederiksen H, Jorgensen N, Andersson AM (2010) Correlations between phthalate metabolites in urine, serum, and seminal plasma from young danish men determined by isotope dilution liquid chromatography tandem mass spectrometry[J]. J Anal Toxicol 34(7):400–410
12. Hines EP, Calatat AM, Silva MJ et al (2009) Concentrations of phthalate metabolites in milk, urine, saliva, and serum of lactating North Carolina women[J]. Environ Health Perspect 117(1):86–92
13. Koch HM, Preuss R, Angerer J (2006) Di(2-ethylhexyl)phthalate (dehp): human metabolism and internal exposure – an update and latest results[J]. Int J Androl 29(1):155–165. discussion 181-155
14. Corrales J, Kristofco LA, Steele WB et al (2015) Global assessment of bisphenol a in the environment: review and analysis of its occurrence and bioaccumulation[J]. Dose-Response Publ Int Hormesis Soc 13(3):1559325815598308
15. Chapin RE, Adams J, Boekelheide K et al (2008) Ntp-cerhr expert panel report on the reproductive and developmental toxicity of bisphenol a[J]. Birth Defects Res B Dev Reprod Toxicol 83(3):157–395
16. Diamanti-Kandarakis E, Bourguignon JP, Giudice LC et al (2009) Endocrine-disrupting chemicals: an endocrine society scientific statement[J]. Endocr Rev 30(4):293–342
17. Asimakopoulos AG, Thomaidis NS, Koupparis MA (2012) Recent trends in biomonitoring of bisphenol a, 4-t-octylphenol, and 4-nonylphenol[J]. Toxicol Lett 210(2):141–154
18. Calafat AM, Longnecker MP, Koch HM et al (2015) Optimal exposure biomarkers for nonpersistent chemicals in environmental epidemiology[J]. Environ Health Perspect 123(7):A166–A168
19. Ye X, Kuklenyik Z, Needham LL et al (2005) Quantification of urinary conjugates of bisphenol a, 2,5-dichlorophenol, and 2-hydroxy-4-methoxybenzophenone in humans by online solid phase extraction-high performance liquid chromatography-tandem mass spectrometry[J]. Anal Bioanal Chem 383(4):638–644
20. Andra SS, Austin C, Yang J et al (2016) Recent advances in simultaneous analysis of bisphenol a and its conjugates in human matrices: exposure biomarker perspectives[J]. Sci Total Environ 572:770–781
21. Liao CY, Kannan K (2012) Determination of free and conjugated forms of bisphenol a in human urine and serum by liquid chromatography-tandem mass spectrometry[J]. Environ Sci Technol 46(9):5003–5009
22. Lacroix MZ, Puel S, Collet SH et al (2011) Simultaneous quantification of bisphenol a and its glucuronide metabolite (bpa-g) in plasma and urine: applicability to toxicokinetic investigations[J]. Talanta 85(4):2053–2059
23. Gerona RR, Woodruff TJ, Dickenson CA et al (2013) Bisphenol-a (bpa), bpa glucuronide, and bpa sulfate in midgestation umbilical cord serum in a northern and Central California population[J]. Environ Sci Technol 47(21):12477–12485

24. Arbuckle TE, Marro L, Davis K et al (2015) Exposure to free and conjugated forms of bisphenol a and triclosan among pregnant women in the mirec cohort[J]. Environ Health Perspect 123(4):277–284
25. Nachman RM, Fox SD, Golden WC et al (2013) Urinary free bisphenol a and bisphenol a-glucuronide concentrations in newborns[J]. J Pediatr 162(4):870–872
26. Battal D, Cok I, Unlusayin I et al (2014) Determination of urinary levels of bisphenol a in a turkish population[J]. Environ Monit Assess 186(12):8443–8452
27. Provencher G, Berube R, Dumas P et al (2014) Determination of bisphenol a, triclosan and their metabolites in human urine using isotope-dilution liquid chromatography-tandem mass spectrometry[J]. J Chromatogr A 1348:97–104
28. Gerona RR, Woodruff TJ, Dickenson CA et al (2013) Bisphenol-a (bpa), bpa glucuronide, and bpa sulfate in midgestation umbilical cord serum in a northern and Central California population[J]. Environ Sci Technol 47(21):12477–12485
29. Waechter J, Thornton C, Markham D et al (2007) Factors affecting the accuracy of bisphenol a and bisphenol a-monoglucuronide estimates in mammalian tissues and urine samples[J]. Toxicol Mech Methods 17(1):13–24
30. Hauck ZZ, Huang K, Li GN et al (2016) Determination of bisphenol a-glucuronide in human urine using ultrahigh-pressure liquid chromatography/tandem mass spectrometry[J]. Rapid Commun Mass Spectrom 30(3):400–406
31. Thuresson K, Hoglund P, Hagmar L et al (2006) Apparent half-lives of hepta- to decabrominated diphenyl ethers in human serum as determined in occupationally exposed workers[J]. Environ Health Perspect 114(2):176–181
32. Valters K, Li H, Alaee M et al (2005) Polybrominated diphenyl ethers and hydroxylated and methoxylated brominated and chlorinated analogues in the plasma of fish from the Detroit river[J]. Environ Sci Technol 39(15):5612–5619
33. Keller JM, Swarthout RF, Carlson BK et al (2009) Comparison of five extraction methods for measuring pcbs, pbdes, organochlorine pesticides, and lipid content in serum[J]. Anal Bioanal Chem 393(2):747–760
34. Lin Y, Feng C, Xu Q et al (2016) A validated method for rapid determination of dibenzo-p-dioxins/furans (pcdd/fs), polybrominated diphenyl ethers (pbdes) and polychlorinated biphenyls (pcbs) in human milk: focus on utility of tandem solid phase extraction (spe) cleanup[J]. Anal Bioanal Chem 408(18):4897–4906
35. Liu LY, He K, Hites RA et al (2016) Hair and nails as noninvasive biomarkers of human exposure to brominated and organophosphate flame retardants[J]. Environ Sci Technol 50(6):3065–3073
36. Ivanisevic J, Zhu ZJ, Plate L et al (2013) Toward omic scale metabolite profiling: a dual separation-mass spectrometry approach for coverage of lipid and central carbon metabolism[J]. Anal Chem 85(14):6876–6884
37. Grigoryan H, Li H, Iavarone AT et al (2012) Cys34 adducts of reactive oxygen species in human serum albumin[J]. Chem Res Toxicol 25(8):1633–1642
38. Da Silva RR, Dorrestein PC, Quinn RA (2015) Illuminating the dark matter in metabolomics[J]. Proc Natl Acad Sci U S A 112(41):12549–12550
39. Zhou B, Xiao JF, Tuli L et al (2012) Lc-ms-based metabolomics[J]. Mol BioSyst 8(2):470–481
40. Rappaport SM, Barupal DK, Wishart D et al (2014) The blood exposome and its role in discovering causes of disease[J]. Environ Health Perspect 122(8):769–774
41. Barr DB, Wang RY, Needham LL (2005) Biologic monitoring of exposure to environmental chemicals throughout the life stages: requirements and issues for consideration for the national children's study[J]. Environ Health Perspect 113(8):1083–1091
42. Lu C, Anderson LC, Morgan MS et al (1998) Salivary concentrations of atrazine reflect free atrazine plasma levels in rats[J]. J Toxicol Environ Health A 53(4):283–292
43. Beane J, Vick J, Schembri F et al (2011) Characterizing the impact of smoking and lung cancer on the airway transcriptome using rna-seq[J]. Cancer Prev Res (Phila) 4(6):803–817

44. Athersuch TJ, Keun HC (2015) Metabolic profiling in human exposome studies[J]. Mutagenesis 30(6):755–762
45. Athersuch T (2016) Metabolome analyses in exposome studies: profiling methods for a vast chemical space[J]. Arch Biochem Biophys 589:177–186
46. Andra SS, Austin C, Patel D et al (2017) Trends in the application of high-resolution mass spectrometry for human biomonitoring: an analytical primer to studying the environmental chemical space of the human exposome[J]. Environ Int 100:32–61
47. Manrai AK, Cui Y, Bushel PR et al (2017) Informatics and data analytics to support exposome-based discovery for public health[J]. Annu Rev Public Health 38:279–294
48. Friedman J, Hastie T, Tibshirani R (2008) Sparse inverse covariance estimation with the graphical lasso[J]. Biostatistics 9(3):432–441
49. Di Q, Rowland S, Koutrakis P et al (2017) A hybrid model for spatially and temporally resolved ozone exposures in the continental United States[J]. J Air Waste Manage Assoc 67(1):39–52
50. Dennis KK, Marder E, Balshaw DM, Cui Y, Lynes MA, Patti GJ, Rappaport SM, Shaughnessy DT, Vrijheid M, Barr DB (2017) Biomonitoring in the era of the exposome[J]. Environ Health Perspect 125:502–510
51. Johns LE, Cooper GS, Galizia A, Meeker JD (2015) Exposure assessment issues in epidemiology studies of phthalates[J]. Environ Int 85:27–39

Effects of Air Pollutants Exposure on Cardiopulmonary and Metabolic Diseases

Bin Xia and Cuiqing Liu

Abstract Instead of a local or regional problem, air pollution is considered as a global issue with potential long-distance atmospheric transportation and rendering health implications. In this chapter, we summarize the evidences from epidemiological and experimental studies on the effects of exposure to particulate matter (PM), ozone (O_3), sulfur dioxide (SO_2), nitrogen oxides (NO_x), and carbon monoxide (CO) on respiratory, cardiovascular, and metabolic systems. Based on in vitro and in vivo studies assessing the effects of air pollutants on cardiopulmonary and metabolic diseases, the biological mechanisms underlying these observations are also summarized. Although both personal intervention and government management and controls have been proved to improve air quality and human health, ongoing research is still needed in this area, with the possibility of therapeutic interventions to reduce the impact of environmental air pollution on cardiopulmonary and metabolic disease in the nearest future.

Keywords Air pollution · Cardiopulmonary disease · Metabolic disease · Intervention

B. Xia
School of Public Health, Fudan University, Shanghai, China

Key Laboratory of Health Technology Assessment, National Health Commission of the People's Republic of China, Fudan University, Shanghai, China

C. Liu (✉)
College of Public Health, Zhejiang Chinese Medical University, Hangzhou, China
e-mail: liucuiqing@zcmu.edu.cn

Y. Zhang (ed.), *Emerging Chemicals and Human Health*,
https://doi.org/10.1007/978-981-32-9535-3_3

1 Introduction

Ambient air pollution is a major global public health problem which affects both developing and developed countries. Nowadays, energy is derived from natural and inexpensive sources for industrial, commercial, and living purposes in many developing countries. However, technologies to mitigate potential air pollution arising from these energy utilizations are limited. For example, refinery emissions are released from different steps of production line, and a typical oil refinery can emit about 50 different atmospheric air pollutants. In addition, some accidental activities such as spillage, venting, decommissioning, and loading of products are also potential sources of gaseous emissions. According to the World Health Organization (WHO) Ambient Air Pollution database which is collected from 1,600 cities in 91 countries, almost 9 out of 10 individuals living in urban areas are affected by air pollution. The WHO has also estimated that each year ambient (outdoor) air pollution causes 3.7 million premature deaths worldwide (http://www.who.int/), with 40,000-50,000 of which occurring in the United States and 300,000 in China. Most of these deaths are from acute and chronic cardiopulmonary events in clinic. Additionally, it has been demonstrated that exposure to air pollution could worsen existing cardiopulmonary diseases and induce metabolic disorder. In this chapter, the current state of knowledge as to the sources of ambient air pollutants and how air pollution is associated with cardiopulmonary and metabolic diseases are discussed. The chapter will mainly discuss evidence from epidemiologic, controlled human, and animal toxicological studies.

2 Main Ambient Air Pollutants

Current air pollution frequently found in urban areas is a dynamic and complex mixture of pollutants from both human-made (anthropogenic) and natural sources. Five main ambient air pollutants are particulate matter (PM), ozone (O_3), sulfur dioxide (SO_2), nitrogen oxides (NO_x), and carbon monoxide (CO). Unlike coarse particles commonly caused by disturbances of crustal materials (dust), fine particles are mainly formed through industrial processes and from traffic-related sources (gasoline and diesel), coal and oil fuel combustion, farming, and road construction. Fine particulate pollution is particularly prevalent in Asia, where biofuel combustion for heating and cooking produces great amounts of fine PM.

Owing to the complexity of its chemical structure and the temporally changing characteristics of the compounds within ambient air, PM is broadly categorized and regulated by aerodynamic diameter (μm). According to the size, PM is commonly subdivided into three classifications: coarse PM with diameters 2.5-10 μm (PM_{10}), fine particles with diameters less than 2.5 μm ($PM_{2.5}$), and ultrafine particles with diameters less than 0.1 μm ($PM_{0.1}$). There are thousands of chemicals and constituents within PM that may solely, or in combination, pose biological hazard. However, the actual "responsible" components remain largely unknown. Nonetheless, it is generally believed that the myriad of compounds that pose intrinsic redox potential (e.g., metals and organic compounds) and/or those capable of activating endogenous sources of oxidative stress within pulmonary tissue (e.g., immune cells) are likely to be the instigators, rather than inert particles (e.g., elemental carbon) [1].

Additionally, with global climate alteration and enlargement of urban centers, it is estimated that ground-level ozone is becoming an even more serious health hazard. Ozone smog forms when NO_X and volatile organic compounds from vehicle, power, and other sources mix with sunlight and heat. Other pollutants (CO, NO_X, and SO_2), mainly produced by fossil fuel combustion and traffic emission, also contribute to air pollution in large urban areas, especially in dense cities in Asia.

Indoor air pollution is also a source of air pollutants, including environmental or second-hand smoke, incomplete combustion of household heating and cooking solid fuels, residential fitment-derived formaldehyde, and volatile organic compounds. Based on the time spent indoors, its contribution to human exposure could be relatively high. In the United States, these primary air contaminants are classified as "criteria" pollutants by the Environmental Protection Agency (EPA), and EPA's National Ambient Air Quality Standards (NAAQS) have been set for each of these pollutants. Actually, these pollutants are present in other countries at different levels. Since these pollutants are closely associated with adverse human health, an extensive literature review should be updated every 5 years to ensure adequate health protection of the public.

3 Association Between Air Pollution and Pulmonary Health Effects

The human respiratory system develops from in utero to adolescence and more than 80% of new alveoli formed in the first 6 years of life following birth. Due to the immature detoxification and metabolic systems, as well as frequent exposure to outdoor air, children are generally more susceptible to airway toxicants than adults. In the elderly, function of particle clearance is usually less efficient or impaired. Thus, old and young populations are more susceptible to inflammation and respiratory complications induced by air pollution.

It has been demonstrated that environmental exposures adversely affected the development of both immune function, lung mechanics, and lung repairment after injury [2]. In addition, those air pollutants-induced adverse effects in the lung include disruption of airway epithelial barrier and cellular signaling pathways, oxidative stress, impairment of phagocytosis, parenchymal destruction, deregulated cell immunity, inflammatory cell infiltration, epigenetic modifications, and autophagy (Fig. 1) [3–10]. These effects are involved in the following respiratory diseases.

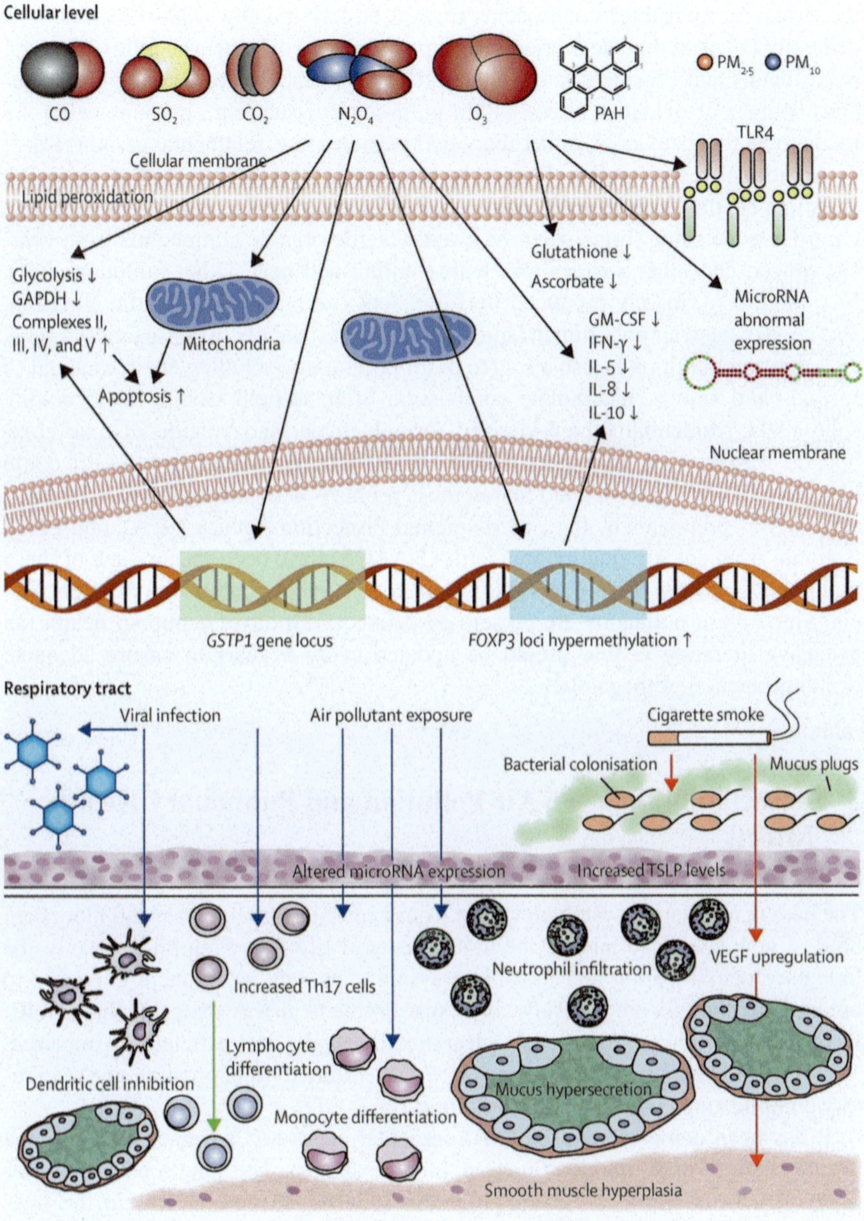

Fig. 1 Working model of air pollutants in the respiratory tract and cells [10]. Abbreviations: *PAH* polycyclic aromatic hydrocarbon, *IFN* interferon, *IL* interleukin, *GM-CSF* granulocyte macrophage-colony stimulating factor, *TSLP* thymic stromal lymphopoietin, *VEGF* vascular endothelial growth factor

3.1 Air Pollution and Asthma

Asthma is characterized clinically by intermittent symptoms of wheezing, dyspnea, and cough with some degree of persistent airflow obstruction. Recently, a series of studies demonstrated that air pollution contributed to increased asthma prevalence and symptom onset. Open fires for cooking are still used in many developing countries or regions (e.g., rural China, India, and Malawi), and coal combustion for heating and cooking conferred higher risks of childhood asthma in China [11]. In addition, indoor air pollution, including environmental tobacco smoke and chemical emissions from new furniture was one of the risk factors for asthma, wheeze and daytime breathlessness amongst pupils [12, 13]. According to a large body of epidemiological evidence, exposure to ambient ozone, nitrogen dioxide, $PM_{2.5}$, and sulphur dioxide was associated with increased asthma exacerbations and hospital admissions [14, 15]. Exposure to $PM_{2.5}$, PM_{10}, ozone, and NO_2 were also found to be associated with a lower forced expiratory volume in the first second (FEV1) and forced vital capacity (FVC) or exacerbated allergic inflammation in children [16–18]. Hua et al. suggested that black carbon conferred greater asthma admission risks than $PM_{2.5}$, with a maximum of a 3-day lag for children in Shanghai, even after stratification by sex, age, and season [19]. These observations were further supported with evidence from mechanism studies and animal experiments. For example, Ozone and SO_2 were shown to cause oxidative stress and affect the airway inflammatory and immune responses in laboratory animals [20, 21]. Moreover, although the relationship between air pollution and adult-onset asthma still remains unclear, a large European study conducted in 23,704 adults from eight countries with a 10-year follow-up further suggested that exposure to traffic-related air pollution containing $PM_{2.5}$ might increase asthma incidence in adults [22].

3.2 Air Pollution and Chronic Obstructive Pulmonary Disease

Air pollution and chornic obstructive pulmonary disease (COPD) is most commonly associated with smoking, but the prevalence of COPD among nonsmokers varies from 1.1% to 40% in different countries [23]. Other risk factors for COPD include occupational exposures and air pollution. Studies demonstrated that indoor/outdoor air pollution and second-hand tobacco smoke contributed to the high incidence of COPD among nonsmokers. In China, development of COPD, especially in rural areas, was attributed to household pollutants, such as coal and wood smoke. The use of improved household stoves in Xuanwei (Yunnan province) resulted in a lower risk of COPD in males (*RR* 0.58; 95% *CI*: 0.49-0.70) and females (*RR* 0.75; 95% *CI*: (0.62-0.92) than that in lifelong unimproved household stove users [24]. In addition, a longitudinal cohort study also demonstrated that improved ventilation along with clean fuels (biogas instead of biomass) usage was associated with lower *OR* (0.28) of development of COPD and slower decline in FEV1 when compared with either single improvement of ventilation or changing clean fuels, or neither [25]. Downs et al. demonstrated that decline in PM_{10} concentration may actually lead to an attenuated impairment in lung function in adult patients [26]. Moreover, data from a global analysis showed that dose–response relationships between air pollutants and COPD mortality were nonlinear and air pollution might induce long-lasting adverse effects on COPD mortality, with mortality risks increasing rapidly at relative lower $PM_{2.5}$ levels (<100 μg/m^3) and reaching a plateau at higher levels (>300 μg/m^3) [27].

3.3 *Air Pollution and Lung Cancer*

Cigarette smoking is a well-known important factor of lung cancer. Recent studies support the role of air pollution in the development of lung cancer. Yu et al. observed somatic genomic mutations in tumor, adjacent normal lung tissues, and peripheral blood samples from 164 non-small cell lung cancer patients from regions with higher levels of air pollutants compared with patients from other regions with lower levels of air pollutants [28]. In addition, air pollution from solid fuel (e.g., coal and wood) combustion predisposes to lung cancer too. An epidemiological study found that a 10 μg/m^3 increase in 2-year average level of household $PM_{2.5}$ derived from coal combustion was correlated with an elevated risk of lung cancer in males (*RR*: 1.06; 95% *CI*: 1.04-1.07) and females (1.15; 1.12-1.18) [29]. Smoky coal is used to be a major source of fuel in rural China such as Xuanwei. A study demonstrated that lung cancer mortality is very high in Xuanwei, indicating smoky coal may confer greater risks than non-smoky coal or other fuels [30].

Based on data from human, animal, and mechanistic studies, the International Agency for Research on Cancer declared outdoor air pollution and related PM as a Group I human carcinogen in October 2013. Consistent with it, a recent study conducted in male drivers aged 30-59 years between 1999 and 2011 found professional drivers exposed to diesel engine exhaust had an elevated risk of lung cancer. The authors suggested that the proportionate cancer incidence ratios for lung cancer in professional drivers increased significantly (1.20, 95% *CI*: 1.13-1.26) during the study period, and that adjustment for the effect of cigarette smoking did not change the associations (1.09, 95% *CI*: 1.03-1.15). The increased risk was consistent throughout study years and age categories [31]. In addition, a meta-analysis showed that ambient exposure to nitrogen oxides, SO_2, and $PM_{2.5}$ from vehicle emissions significantly increased the risk of lung cancer [32]. Another study indicates that polycyclic aromatic hydrocarbons (PAH) derived from smoky coal burning may also be related with lung cancer [30]. The potential mechanism of lung carcinogenesis remains further study.

4 Association Between Air Pollution and Cardiovascular Disease

The cardiovascular or circulatory system delivers blood to organs and tissues. It consists of the heart, arteries, veins, and capillaries. Cardiovascular disease (CVD) is the number one cause of death in developed nations and is the leading cause of years of life lost due to morbidity and mortality globally [33]. Vascular dysfunction, diabetes, physical inactivity, and smoking have been well described as risk factors for CVDs. Recently, association of air pollution exposure with CVD aroused emerging attention [34]. Two landmark cohort-based mortality studies, the Harvard Six Cities and the ACS (American Cancer Study) studies, reported in the mid-1990s demonstrated that air pollutants (e.g., $PM_{2.5}$ and sulfate) were associated with increase in all-cause and cardiopulmonary diseases. In both studies, PM air pollution-related mortality was substantially higher for cardiovascular than for pulmonary-related diseases [35, 36]. As air pollution is a modifiable risk factor, elucidation of its adverse cardiovascular effects would provide more evidence for the government to adopt a policy on reducing air pollution levels and providing health care.

4.1 PM and CVD

Due to the small size, $PM_{2.5}$ and $PM_{0.1}$ are inhaled deeply into the lungs, followed by depositing in the alveoli and entering the pulmonary and systemic circulation. Thus, the role of PM as a risk factor to CVD has been well established. Scientific studies have linked increases in daily $PM_{2.5}$ exposure with increased cardiovascular morbidity and mortality. Exposure to ambient air-borne $PM_{2.5}$ is associated with increased incidences of specific acute CVDs, such as cardiac arrhythmia, atrial fibrillation, ischemic stroke, heart failure, myocardial infarction, as well as peripheral arterial and venous diseases. Multiple studies also have found the significant associations between $PM_{2.5}$ exposure and elevated blood pressure or decreased heart rate variability (HRV) [37]. However, the relationship between PM_{10} and CVD is still controversial, for studies demonstrated that after adjustment for $PM_{2.5}$, there were no statistically significant associations between coarse PM and hospital admissions for CVD or changes in HRV [38, 39]. Table 1 summarizes representative cardiovascular effects of PM in some largest short-term or long-term studies.

Potential mechanisms involved in PM-mediated CVD progression include systemic inflammation, oxidative stress, increased blood coagulability, alterations in autonomic balance, and vascular dysfunction [50]. First, pollutants may induce autonomic nervous system (ANS) imbalance favoring sympathetic over parasympathetic tone, which may be prompted by lung irritant sensory receptors and afferent nerve stimulation. Second, air pollution could cause the release of vasculo-active molecules [e.g., histamine or endothelin] or endogenous proinflammatory mediators

Table 1 Summary of representative studies on short-term or long-term cardiovascular effects of PM air pollution

Publications	Population	Key findings
Short-term cardiovascular effects		
Dominici et al. [40]	Time-series study on 11.5 million U.S. individuals aged over 65 years	Increase of 1.3% in the risk for CHF for each 10 μg/m³ $PM_{2.5}$ elevation
Analitis et al. [41]	Time-series study conducted in 29 European cities on 43 million adults	Per 20 μg/m³ PM_{10} elevation was associated with increase of 1.5% daily cardiovascular deaths
Dominici et al. [42]	Time-series study including 50 million adults from 20 U.S. cities	Increase of 0.6% of daily cardiorespiratory deaths for each 20 μg/m³ PM_{10} elevation
Zanobetti et al. [43]	631 repeated visits for cardiac rehabilitation from 62 Boston residents with cardiovascular disease	Per 10-μg/m³ $PM_{2.5}$ elevation during previous 5 days was associated with 2.8 mm hg increase in mean arterial blood pressure
Pope et al. [44]	Case cross-over study on 12,865 U.S. patients undergoing coronary arteriography	Per 10-μg/m³ elevation of $PM_{2.5}$ was associated with increase of 4.5% daily rate of acute ischemic coronary events
Long-term cardiovascular effects		
Laden et al. [45]	Prospective cohort study on 8111 people from 6 U.S. cities followed for 28 years	Per 10 μg/m³ elevation of $PM_{2.5}$ was associated with increase in the relative risk for cardiovascular death (*RR*, 1.28; 95% *CI*, 1.13-1.44)
Pope et al. [46]	Prospective cohort study on 500,000 residents of 150 U.S. cities followed for 16 years	Per 10 μg/m³ elevation in long-term $PM_{2.5}$ exposure was associated with 12% and 18% increase in the risk for death from cardiovascular causes and ischemic heart disease, respectively
Miller et al. [47]	Prospective cohort study on 65,893 postmenopausal women in 36 U.S. cities followed for 6 years	Per 10 μg/m³ elevation in long-term $PM_{2.5}$ exposure was associated with 24% and 76% increase in the risk for cardiovascular events and cardiovascular mortality, respectively
Toren et al. [48]	Prospective cohort study on 71,778 unexposed male construction workers and 176,309 PM exposed male construction workers	Occupational exposure to particulate air pollution with occupational exposure was associated with increase in the relative risk for ischemic heart disease (*RR*, 1.13; 95% *CI*, 1.07-1.19)
Dadvand et al. [49]	Prospective cohort study on 8,398 pregnancies in Barcelona, Spain	An IQR increase in $PM_{2.5}$ during the entire pregnancy (5.1 μg/m³) and the third trimester (7.3 μg/m³) was associated with preeclampsia: *ORs* of 1.3 (95% *CI*: 1.0-1.7) and 1.5 (95% *CI*: 1.1-2.0), respectively. An IQR (2.1×10^{-5}/m) increase in $PM_{2.5}$ absorbance was associated with preeclampsia in the third trimester, *OR* of 1.4 (95% *CI*: 1.0-1.9)

(e.g., cytokines IL-6 or IL-1) from lung cells and activated immune cells that have been shown to "spill-over" into the systemic circulation, resulting in detrimental vascular effects. Third, some soluble constituents such as metals or nano-sized particles could translocate across the alveolar membrane and gain access to the bloodstream and directly influence the vascular endothelium. The relevance of each pathway may vary depending upon susceptibility of the individuals and pollutants characteristics, and these pathways are likely to overlap to some degree in their adverse actions. In sum, air pollution can induce cardiovascular detriment in a biphasic manner consisting of an initial response within minutes-to-hours (due to acute ANS imbalance) and a subsequent arterial vasoconstrictor responsiveness via endothelial dysfunction, oxidative stress, and inflammation [50].

4.2 O_3 and Cardiovascular Diseases

O_3 is the primary oxidant of concern in photochemical smog due to its inherent bioreactivity. It is a bluish, explosive, irritating, and highly toxic gas. So far, effects of ground-level O_3 on the CVD are still controversial. Although no associations were observed in studies conducted in Tucson, Arizona, London, the United Kingdom, Edinburgh, and Scotland, a positive association between O_3 and ischemic heart disease was observed in Helsinki, Finland [51–53]. O_3 has a high oxidizing power by reacting with biomolecules to form ozonides and free radicals. This process triggers an inflammatory response in the body, which induces adverse effects in both pulmonary and cardiovascular systems. According to Srebot et al., pulmonary oxidant stress mediated by PM and/or O_3 exposure resulted in downstream perturbations in the cardiovasculature, as the pulmonary and cardiovascular systems were intricately associated [54]. Mechanisms by which ground-level O_3 exposure induced cardiotoxicity have been explored, which include modification of endothelial function, vascular angiotasis, alterations in autonomic control of cardiac frequency, activation of systemic inflammatory response mediated by cytokines, and increase of oxidative stress. In a randomized, double-blind, crossover chamber study by Brook et al., O_3 caused acute arterial vasoconstriction in healthy adults with altered macrovascular diameter and tone [55]. Ambient levels of O_3 pollution may lead to short-term autonomic imbalance, as evidenced by reduced high-frequency component of HRV, with patients in hypertension being particularly susceptible to this effect [53]. In addition, stimulation of peripheral human blood mononuclear cells with O_3 induces increase in lipid peroxidation and protein thiol group content.

4.3 NO_2 and CVD

NO_2 is a gaseous chemical and a member of nitrogen oxides. The harmful health effects of NO_2 may potentially result from NO_2 itself or its reaction products such as O_3. Previous studies have reported that exposure to higher concentration of NO_2 was associated with daily hospital emergency for ischemic heart diseases, as well as for subsequent cardiac insufficiency and arrhythmia. For instance, Mar et al. has reported that an interquartile range increase in NO_2 was associated with an increase of 6.1% of cardiovascular mortality [56]. The cardiovascular effects of NO_2 were mainly observed in patients with CVDs aged 65 years or older who had high risks of atherogenesis [57]. Additionally, Takano et al. found that daily exposure to ambient level (0.16 ppm) of NO_2 could enhance atherogenic lipid metabolisms primarily in the Otsuka Long-Evans Tokushima Fatty (OLETF) rats, but less in the Long-Evans Tokushima Otsuka (LETO) rats, which suggested that ambient NO_2 air pollution was an atherogenic risk primarily in obese subjects [58].

4.4 SO_2 and CVD

SO_2 is a colorless, highly soluble, and chemical irritant. Interestingly, SO_2 is a two-edged sword. In one hand, SO_2 is an endogenous gaseous signaling molecule involved in the regulation of cardiovascular functions [59]. In the other hand, SO_2 is considered to be toxic and detrimental to human health. For example, previous study has suggested that daily exposure to SO_2 pollution might play an independent role in triggering ischemic cardiac events in Europe [60]. Blood viscosity is a risk factor of CVD. A positive association of CVD incidence with increased levels of ambient total suspended particles and SO_2 was observed as well [61].

4.5 CO and CVD

CO is an odorless gas that causes thousands of deaths each year in North America. Evidence from human studies has shown that CO typically affects oxygenation of tissue by production of carboxyhemoglobin (COHb), which subsequently results in adverse cardiovascular effects, such as exacerbating the ischemic heart disease. The harmful effects of CO are more profound in the myocardium than in peripheral tissues due to the high oxygen extraction by the myocardium at rest [62]. In addition, exposure to CO has been implicated in the process of atherosclerosis.

5 Other Air Pollutants and CVD

Other air pollutants, including cigarette smoke, biomass smoke, and lead pollution were also reported to contribute to the development of CVD. As a well-established risk factor for CVD, cigarette smoking is associated with a number of clinical atherosclerotic syndromes, including sudden death, acute coronary syndromes, stable angina, and stroke. Smokers who inhale deeply are more likely to have an increased risk of both abdominal aortic aneurysm and symptomatic peripheral arterial disease. Household burning of solid fuels releases several pollutants including inhalable PM, PAHs, heavy metals, and many other organic pollutants, which have been linked to CVDs [63]. The cardiovascular effects of lead exposure include elevation of blood pressure, hypertension, coronary heart disease, stroke, and peripheral arterial disease, as well as other cardiovascular function abnormalities including left ventricular hypertrophy and alterations in cardiac rhythm [64].

6 Association Between Air Pollution and Metabolic Diseases

6.1 Air Pollution and Diabetes Mellitus (DM)

Type 2 diabetes mellitus (T2DM) affects approximately 9% of the global population and accounts for 2.7% of global deaths (WHO, 2012), and the number is expected to increase to 592 million by 2035 [65], with the sharp increases occurring in low- and middle-income countries. Epidemiologic studies have uncovered a number of factors for the risk of metabolic disorders, including changes in lifestyle factors due to urbanization, such as reduced physical activity, overnutrition, lack of sleep, and alterations in light cycle. Recently, there is cumulative evidence of the association between air pollution exposure and T2DM. $PM_{2.5}$ and other air pollutants exaggerate insulin resistance and the development of T2DM. Table 2 summarizes a number of representative studies concerning a positive association between long-term ambient air pollution exposure and increased risk of T2DM.

- *Gestational diabetes*

Women who experience gestational diabetes are at a higher risk for subsequent T2DM. Previous studies have demonstrated that air pollutants might have the potential to affect gestational diabetes development too. A large retrospective cohort including 219,952 singleton deliveries showed exposure to higher levels of NOx and SO_2 during the 3 months preconception was related to higher relative risks for gestational diabetes in women [76]. The results were consistent when considering

Table 2 Epidemiological associations among air pollutants, diabetes, and diabetes-associated mortality

Publications	Location and subject number	Main pollutants	Principal findings
Brook et al. [66]	Ontario, Canada; 7,634 subjects	NO_2	Per 1 ppb elevation of NO_2 was associated with increase in DM prevalence (*OR*, 1.04; 95% *CI*: 1.00-1.08) in women. NO significant association in men
Kramer et al. [67]	Ruhr, Germany; 1,775 nondiabetic women	PM_{10} (mean: 47 μg/m^3); NO_2	Per IQR increase in PM_{10}, NO_2 or traffic exposure over mean 16 years was associated with increases in adjusted HR for DM development (ranging from 1.15 to 1.42)
Pearson et al. [68]	USA; subject number was estimated by county level	$PM_{2.5}$	DM prevalence associated with $PM_{2.5}$ at county level (1% increase per 10 μg/m^3) and persistently associated with $PM_{2.5}$ at counties below annual standards ($PM_{2.5}$ <15 μg/m^3).
Dijkema et al. [69]	The Netherlands; 8,018 subjects including 619 DM subjects	NO_2	DM prevalence increased with exposure within traffic in a 250 m buffer when compared to the lowest quartile although nonsignificant
Coogan et al. [70]	Los Angeles, USA; 3,992 women	NO_2 and $PM_{2.5}$	NO_2 exposure associated with 10-year incidence of DM (adjusted *HR*, 1.25; 95% *CI*: 1.07-1.46). $PM_{2.5}$ not related.
Andersen et al. [71]	Denmark; 51,818 nondiabetic participants	NO_2 level (IQR: 4.9 (μg/m^3)	NO_2 exposure positively associated with confirmed DM (1.04; 95% *CI*: 1.00-1.08) in nonsmokers and active people.
Brook et al. [72]	Canada; 2.1 million adults	Mean $PM_{2.5}$ (8.7 μg/m^3)	The adjusted HR of diabetes-associated mortality for a 10 μg/m^3 $PM_{2.5}$ was 1.49 (95% *CI*: 1.37-1.62).
Dales et al. [73]	Montreal, Canada; 158,350 elderly adults	$PM_{2.5}$ (mean 8.74 μg/m^3), NO_2 (mean 37.90 μg/m^3), O_3, CO and SO_2	Positive associations were found between daily non-accidental mortality and all air pollutants but O_3.
Eze et al. [74]	Switzerland; 6,392 subjects	NO_2 and PM_{10}	Per 10 μg/m^3 increase in NO_2 and PM_{10} exposure were associated with increasing diabete prevalence (*OR*: 1.19, 95% *CI*: 1.03, 1.38) and (*OR*: 1.40, 95% *CI*: 1.17-1.67)
Zanobetti et al. [75]	USA; 121 communities	$PM_{2.5}$	Short-term exposure to $PM_{2.5}$ (per 10 μg/m^3 increase in 2-day average) was associated with an increase in hospitalization risks for diabetes (1.14%, 95% *CI*: 0.56, 1.73) and all-cause mortality risk (0.64%, 95% *CI*: 0.42, 0.85).

only air pollution exposures that occurred during the first trimester. Consistent with it, a dose–response relationship was also found in a Swedish study linking registry data from 81,110 pregnancies with individual NOx exposure [77]. However, a study conducted in Taiwan demonstrated that exposure to O_3 might lead to more preterm births in women with gestational diabetes [78]. With O_3 air pollution getting more serious, further exploring the effects of O_3 exposure on maternal metabolism and offspring are necessary.

- *Precursors of T2DM*

Exposure to ambient air pollution is also associated with predisposing conditions for diabetes. In 9,102 newly diagnosed T2DM patients in Germany, higher regional levels of PM_{10} were associated with higher HbA1c concentrations, a biomarker of the average blood glucose levels in an individual during the initial 30-120 days [79]. Adverse effects of several pollutants on HbA1c and fasting glucose concentrations were also observed in an elderly population in Taiwan [80]. Insulin resistance is another potential precursor of T2DM, which could lead to elevated fasting glucose accompanied by elevated insulin levels due to impaired insulin action. Previous study conducted among 560 elderly individuals in Korea showed that short-term exposure to NO_2 was associated with elevated homeostatic model assessment of insulin resistance (HOMA-IR) in participants without a history of diabetes. For participants with diabetes, especially for those carrying risk genotypes in the oxidative stress-related human glutathione S-transferase genes (*GSTM1*, *GSTT1*, and *GSTP1*), PM_{10}, NO_2, and O_3 were all associated with increased HOMA-IR [81]. In addition, long-term exposure to NO_2 and PM_{10} or short-term exposure to PM_{10} and CO were associated with elevated HOMA-IR in healthy children aged from 10 to 18 years [82], with increased inflammation and oxidized low-density lipoprotein.

6.2 Air Pollution and Obesity

Obesity is becoming a major global public health problem and consumes substantial social resources. The prevalence of obesity more than doubled among the American adults from 1976-1980 to 2013-2014. The global obesity prevalence is expected to exceed 21% in women and reach 18% in men by 2025 [83].

Interestingly, several mechanisms may link unhealthy body weight to air pollution, which is likely to be highly complex and differ across subpopulation (e.g., adults versus children), severity, and type of pollution. Air pollution could increase oxidative stress and adipose tissue inflammation, induce hepatic lipid accumulation, and decrease glucose utilization in skeletal muscle, which may result in metabolic dysfunction [84, 85]. In addition, air pollution may indirectly prevent people from engaging in regular physical activity by inducing cardiovascular and respiratory symptoms (e.g., decreased lung function and elevated blood pressure) and impairing exercise performance. Emerging evidence demonstrates that air pollution level is negatively associated with physical activity and/or positively associated with leisure-time physical inactivity. A unit (μg/m^3) increase in ambient $PM_{2.5}$ concentration was associated with an increase in the odds of physical inactivity by 1.1% among the U.S. adults [86]. Participants, in particular those with respiratory diseases, self-reported a reduction in outdoor activities to mitigate the detrimental impact of air pollution. Indeed, smog appearance could also discourage people from engaging in outdoor activities [87].

6.3 *Mechanistic Insights into Air Pollution and Diabetes*

A number of mechanisms have been proposed for extrapulmonary effects of inhaled pollutants. On the basis of experimental studies examining the metabolic effects of ambient PM in mouse models, studies suggested that systemic inflammation, oxidative stress, ER stress, and neuronal mechanisms might be involved in insulin-sensitive organs [88, 89].

In the visceral adipose tissue, $PM_{2.5}$ exposure might induce shift of adipose tissue macrophages to a pro-inflammatory phenotype (M1) characterized by increase in TNF-a and IL-6, and decrease in IL-10 gene expression [90]. In the same study, a transgenic model of yellow-fluorescent protein expression restricted to monocytes (c-fms^{YFP}) rendered insulin resistant with high-fat diet, and exposure to $PM_{2.5}$ could result in doubling in the number of endothelial adherent YFP$^+$ cells in mesenteric fat with a sixfold increase in monocytes within adipose [90]. Thus, PM mediates migration and adhesion of YFP$^+$ cells in visceral fat depots. In another study, a CCR2 KO model was exposed to $PM_{2.5}$, increased number of immune cells [F4/80$^+$ (macrophage) or F4/80$^+$/CD11c$^+$ (M1 macrophage)] in visceral adipose tissue in response to $PM_{2.5}$ exposure was ablated by CCR2 deficiency, suggesting a dependence of $PM_{2.5}$ on CCR2 in recruitment of innate immune cells [85].

The central nervous system (CNS) plays a critical role in energy balance with multiple environmental and internal signals serving as cues to trigger the requisite behavioral and physiological response to maintain energy homeostasis. Interestingly, $PM_{2.5}$ exposure could induce hyperglycemia and insulin resistance, which is accompanied by increased hypothalamic IL-6, IKKβ, and TNFα mRNA expression and microglial/astrocyte reactivity. Targeting the NF-κB pathway with intracerebroventricular administration of an IKKβ inhibitor improved insulin sensitivity, glucose tolerance, reduced peripheral inflammation, and rectified energy homeostasis in response to $PM_{2.5}$ [91]. Intra-cerebroventricular administration of IMD-0354 (an IKKβ inhibitor) inhibited the expression of the enzymes for gluconeogenesis and lipogenesis in the liver [92]. Thus, hypothalamic inflammation plays a key role in $PM_{2.5}$ exposure-mediated peripheral inflammation, exaggeration of type II diabetes, hepatic glucose, and lipid metabolism disorder.

How may air pollutants induce CNS inflammation and secondarily lead to systemic metabolic disorder? For the most widely studied air pollutants, evidence supports the hypothesis that exposure to PM components of air pollution, ozone, or other gas upon inhalation can oxidatively modify surfactant lining and cellular com-

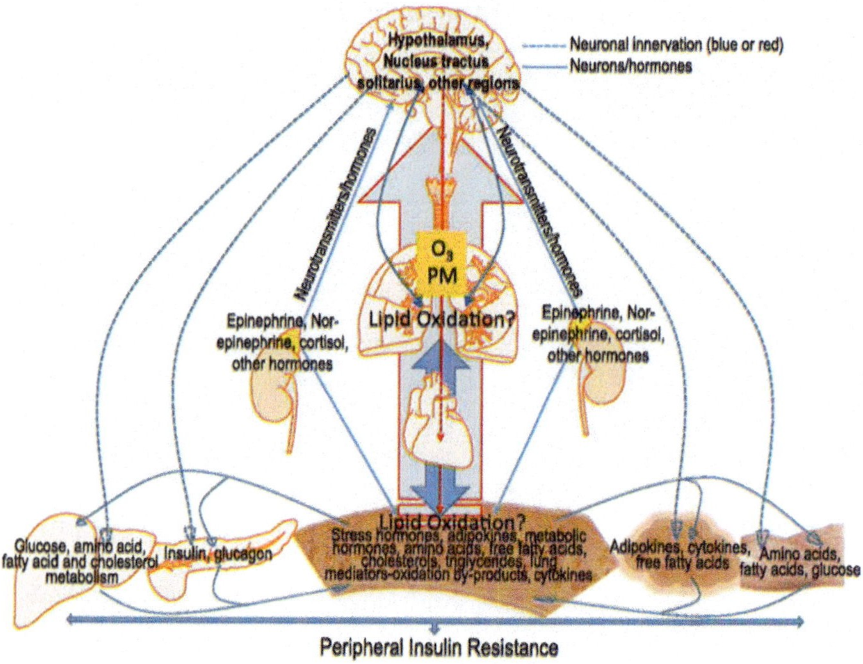

Fig. 2 Potential mechanisms involved in air pollution-induced insulin resistance and diabetes [89]

ponents. Such exposure can also induce stress response to injury through direct sympathetic stimulation, activation of hypothalamus–pituitary–adrenal axis, and subsequently releasing of hormones from the adrenal cortex. These processes may trigger a release of free fatty acids, glucose, and branched-chain amino acids into the circulation (Fig. 2). A variety of systemic inflammatory and homeostatic metabolic mechanisms might be involved in neurohormone regulation, and these factors are involved in insulin resistance, dyslipidemia, and glycemia.

7 Targets for Intervention

Instead of a local or regional problem, air pollution now is recognized as a global issue with potential long-distance atmospheric transportation and rendering health implications. Most importantly, there is no threshold for exposure to particulate air pollution below which exposure is safe. It indicates that individuals are always at risk for adverse events, whether they reside in polluted areas or communities in compliance with EPA's NAAQS. Therefore, apart from strengthening efforts to guarantee air quality compliance with the NAAQS, we might take effective interventions to mitigate the adverse health outcomes of exposure to air pollution for susceptible populations living in areas with high pollution. For instance, personal

Table 3 Personal- and organizational-level interventions to reduce exposures or susceptibility to air pollution

Intervention type	Examples
Air purifiers	HEPA designated filters remove PM at 0.3 microns in diameter by 99.97%; set indoor HEPA filter systems in living spaces and bedrooms could be helpful
Facemasks	N95 designated respirators remove at least 95% of PM at 0.3 microns
Reduce in-traffic exposures	In-cabin HEPA filters, closing car windows, and recirculating in-cabin air
Reduce traffic emissions	Diesel particle traps, catalytic converters, and alternative fuels (natural gas, electric cars)
Reduce in-home penetrance	Closing windows, using central air conditioning filtration, and improving home efficiency
Cleaner cook-stoves	Using cleaner burning liquid fuel stoves and improving stove emission ventilation
Lifestyle changes	Awareness of health risks of travel to heavily polluted regions; Avoiding point sources (roadways), exercising indoors during peak exposure time (rush hour)
Medications and diet	Maintaining healthy diet (vegetables), taking antioxidants, fish oil, and evidence-based medications (statins)
Municipality and city changes	Setting green spaces, natural barriers from roadways and improving urban design (bike paths >400 m from roadways)

indoor PM exposure can be minimized by using particulate air filters, air conditioning, avoiding use of indoor combustion for heating and cooking, and smoking cessation. Susceptible groups may benefit from limiting their outdoor exercise during poor air quality days. Since air pollution remains a complex mixture of anthropogenic pollutants and natural sources, it is essential that local, national, and global efforts are undertaken by government, enterprises, and individuals to lessen the burden of air pollution and provide better health protection. Effective measures to reduce exposure to air pollution, both at organizational and personal level (Table 3), could lead to substantial healthy benefit.

8 Conclusion

Growing bodies of evidence have demonstrated modest increase in risk of cardiopulmonary and metabolic diseases associated with exposure to major air pollutants. The increase in the relative risk for these adverse health outcomes due to air pollution for an individual is relatively small compared with other established risk factors, such as diet, physical activity, and behavior. Given the ubiquitous nature of air pollution and the economic costs of cardiopulmonary and obesity/DM-related diseases, even conservative risk estimates would still translate into a substantial increase in the population attributable fraction of public health problem related to air pollutants. Additional research is therefore needed to confirm the currently available evidence and to explore the impact of air-borne pollutants in populous cities of developing countries. Further investigations on effective interventions that improve air quality and attenuate the adverse effects are therefore warranted.

References

1. Mills NL, Donaldson K, Hadoke PW et al (2009) Adverse cardiovascular effects of air pollution [J]. Nat Clin Pract Cardiovasc Med 6(1):36–44
2. Martino D, Prescott S (2011) Epigenetics and prenatal influences on asthma and allergic airways disease [J]. Chest 139(3):640–647
3. Bahadar H, Mostafalou S, Abdollahi M (2014) Current understandings and perspectives on non-cancer health effects of benzene: a global concern [J]. Toxicol Appl Pharmacol 276(2):83–94
4. Wang ZL (2013) Association between chronic obstructive pulmonary disease and lung cancer: the missing link [J]. Chin Med J 126(1):154–165
5. Xing YF, Xu YH, Shi MH et al (2016) The impact of pm2.5 on the human respiratory system [J]. J Thorac Dis 8(1):E69–E74
6. Huang SK, Zhang Q, Qiu Z et al (2015) Mechanistic impact of outdoor air pollution on asthma and allergic diseases [J]. J Thorac Dis 7(1):23–33
7. Zhang QL, Qiu ZM, Chung KF et al (2015) Link between environmental air pollution and allergic asthma: east meets west [J]. J Thorac Dis 7(1):14–22
8. Wei J, Li F, Yang J et al (2015) Micrornas as regulators of airborne pollution-induced lung inflammation and carcinogenesis [J]. Arch Toxicol 89(5):677–685
9. Chen ZH, Wu YF, Wang PL et al (2016) Autophagy is essential for ultrafine particle-induced inflammation and mucus hyperproduction in airway epithelium [J]. Autophagy 12(2):297–311
10. Guan WJ, Zheng XY, Chung KF et al (2016) Impact of air pollution on the burden of chronic respiratory diseases in China: time for urgent action [J]. Lancet 388(10054):1939–1951
11. Zheng TZ, Niu SR, Lu BY et al (2002) Childhood asthma in Beijing, China: a population-based case-control study [J]. Am J Epidemiol 156(10):977–983
12. Ma YN, Zhao Y, Liu YQ et al (2013) [effects of indoor air pollution on asthma and asthma-related symptoms among children in Shenyang city] [J]. Zhonghua Yu Fang Yi Xue Za Zhi Chin J Prevent Med 47(1):49–54
13. Zhao Z, Zhang Z, Wang Z et al (2008) Asthmatic symptoms among pupils in relation to winter indoor and outdoor air pollution in schools in Taiyuan, China [J]. Environ Health Perspect 116(1):90–97
14. Cheng MH, Chen CC, Chiu HF et al (2014) Fine particulate air pollution and hospital admissions for asthma: a case-crossover study in Taipei [J]. J Toxicol Environ Health-Part A-Curr Issues 77(18):1075–1083
15. Ko FWS, Tam W, Wong TW et al (2007) Effects of air pollution on asthma hospitalization rates in different age groups in Hong Kong [J]. Clin Exp Allergy 37(9):1312–1319
16. Rice MB, Ljungman PL, Wilker EH et al (2013) Short-term exposure to air pollution and lung function in the Framingham heart study [J]. Am J Respir Crit Care Med 188(11):1351–1357
17. Molter A, Agius RM, De Vocht F et al (2013) Long-term exposure to pm10 and no2 in association with lung volume and airway resistance in the Maas birth cohort [J]. Environ Health Perspect 121(10):1232–1238
18. Carosino CM, Bein KJ, Plummer LE et al (2015) Allergic airway inflammation is differentially exacerbated by daytime and nighttime ultrafine and submicron fine ambient particles: Heme oxygenase-1 as an indicator of pm-mediated allergic inflammation [J]. J Toxicol Environ Health-Part a-Curr Issues 78(4):254–266

19. Hua J, Yin Y, Peng L et al (2014) Acute effects of black carbon and pm2.5 on children asthma admissions: a time-series study in a Chinese city [J]. Sci Total Environ 481:433–438
20. Wiegman CH, Li F, Clarke CJ et al (2014) A comprehensive analysis of oxidative stress in the ozone-induced lung inflammation mouse model [J]. Clin Sci 126(5–6):425–440
21. Li RJ, Kou XJ, Tian JJ et al (2014) Effect of sulfur dioxide on inflammatory and immune regulation in asthmatic rats [J]. Chemosphere 112:296–304
22. Jacquemin B, Siroux V, Sanchez M et al (2015) Ambient air pollution and adult asthma incidence in six european cohorts (escape) [J]. Environ Health Perspect 123(6):613–621
23. Bang KM (2015) Chronic obstructive pulmonary disease in nonsmokers by occupation and exposure: a brief review [J]. Curr Opin Pulm Med 21(2):149–154
24. Song QK, Christiani DC, Wang XR et al (2014) The global contribution of outdoor air pollution to the incidence, prevalence, mortality and hospital admission for chronic obstructive pulmonary disease: a systematic review and meta-analysis [J]. Int J Environ Res Public Health 11(11):11822–11832
25. Zhou YM, Zou YM, Li XC et al (2014) Lung function and incidence of chronic obstructive pulmonary disease after improved cooking fuels and kitchen ventilation: A 9-year prospective cohort study [J]. PLoS Med 11(3):e1001621
26. Downs SH, Schindler C, Liu LJS et al (2007) Reduced exposure to pm10 and attenuated age-related decline in lung function [J]. N Engl J Med 357(23):2338–2347
27. Burnett RT, Pope CA, Ezzati M et al (2014) An integrated risk function for estimating the global burden of disease attributable to ambient fine particulate matter exposure [J]. Environ Health Perspect 122(4):397–403
28. Yu XJ, Yang MJ, Zhoua B et al (2015) Characterization of somatic mutations in air pollution-related lung cancer [J]. EBioMedicine 2(6):583–590
29. Guo YM, Zeng HM, Zheng RS et al (2016) The association between lung cancer incidence and ambient air pollution in China: a spatiotemporal analysis [J]. Environ Res 144:60–65
30. Downward GS, Hu W, Rothman N et al (2014) Polycyclic aromatic hydrocarbon exposure in household air pollution from solid fuel combustion among the female population of Xuanwei and fuyuan counties, China [J]. Environ Sci Technol 48(24):14632–14641
31. Koh DH, Kong HJ, Oh CM et al (2015) Lung cancer risk in professional drivers in Korea: a population-based proportionate cancer incidence ratio study [J]. J Occup Health 57(4):324–330
32. Zhang WS, Li FT, Gao WY (2017) Traffic-related air pollution and lung cancer: a meta-analysis [J]. Thoracic Cancer 8(5):546–546
33. Incidence GDI (2017) Global, regional, and national age-sex specific mortality for 264 causes of death, 1980-2016: a systematic analysis for the global burden of disease study 2016 (vol 390, pg 1211, 2017) [J]. Lancet 390(10106):E38–E38
34. Sidney S, Quesenberry CP, Jaffe MG et al (2016) Recent trends in cardiovascular mortality in the United States and public health goals [J]. JAMA Cardiol 1(5):594–599
35. Dockery DW, Pope CA 3rd, Xu X et al (1993) An association between air pollution and mortality in six U.S. cities [J]. N Engl J Med 329(24):1753–1759
36. Pope CA 3rd, Thun MJ, Namboodiri MM et al (1995) Particulate air pollution as a predictor of mortality in a prospective study of U.S. adults [J]. Am J Respir Crit Care Med 151(3 Pt 1):669–674
37. Grahame TJ (2009) Does improved exposure information for pm2.5 constituents explain differing results among epidemiological studies? [J]. Inhal Toxicol 21(5):381–393
38. Peng RD, Chang HH, Bell ML et al (2008) Coarse particulate matter air pollution and hospital admissions for cardiovascular and respiratory diseases among medicare patients [J]. Jama-J Am Med Assoc 299(18):2172–2179
39. Liao DP, Creason J, Shy C et al (1999) Daily variation of particulate air pollution and poor cardiac autonomic control in the elderly [J]. Environ Health Perspect 107(7):521–525
40. Dominici F, Peng RD, Bell ML et al (2006) Fine particulate air pollution and hospital admission for cardiovascular and respiratory diseases [J]. JAMA 295(10):1127–1134

41. Analitis A, Katsouyanni K, Dimakopoulou K et al (2016) Short-term effects of ambient particles on cardiovascular and respiratory mortality [J]. Epidemiology 17(2):230–233
42. Dominici F, Mcdermott A, Daniels M et al (2015) Revised analyses of the national morbidity, mortality, and air pollution study: mortality among residents of 90 cities [J]. J Toxicol Environ Health Part A 68(13–14):1071–1092
43. Zanobetti A, Canner MJ, Stone PH et al (2004) Ambient pollution and blood pressure in cardiac rehabilitation patients [J]. Circulation 110(15):2184–2189
44. Pope CA 3rd, Muhlestein JB, May HT et al (2006) Ischemic heart disease events triggered by short-term exposure to fine particulate air pollution [J]. Circulation 114(23):2443–2448
45. Laden F, Schwartz J, Speizer FE et al (2016) Reduction in fine particulate air pollution and mortality – extended follow-up of the Harvard six cities study [J]. Am J Respir Crit Care Med 173(6):667–672
46. Pope CA, Burnett RT, Thurston GD et al (2004) Cardiovascular mortality and long-term exposure to particulate air pollution – epidemiological evidence of general pathophysiological pathways of disease [J]. Circulation 109(1):71–77
47. Miller KA, Siscovick DS, Sheppard L et al (2007) Long-term exposure to air pollution and incidence of cardiovascular events in women [J]. N Engl J Med 356(5):447–458
48. Toren K, Bergdahl IA, Nilsson T et al (2017) Occupational exposure to particulate air pollution and mortality due to ischaemic heart disease and cerebrovascular disease [J]. Occup Environ Med 64(8):515–519
49. Dadvand P, Figueras F, Basagana X et al (2013) Ambient air pollution and preeclampsia: a spatiotemporal analysis [J]. Environ Health Perspect 121(11–12):1365–1371
50. Brook RD (2008) Cardiovascular effects of air pollution [J]. Clin Sci (London, England: 1979) 115(6):175–187
51. Ponka A, Virtanen M (1996) Low-level air pollution and hospital admissions for cardiac and cerebrovascular diseases in Helsinki [J]. Am J Public Health 86(9):1273–1280
52. Prescott GJ, Cohen GR, Elton RA et al (1998) Urban air pollution and cardiopulmonary ill health: a 14.5 year time series study [J]. Occup Environ Med 55(10):697–704
53. Holguin F, Tellez-Rojo MM, Hernandez M et al (2003) Air pollution and heart rate variability among the elderly in Mexico City [J]. Epidemiology 14(5):521–527
54. Srebot V, Gianicolo EA, Rainaldi G et al (2009) Ozone and cardiovascular injury [J]. Cardiovasc Ultrasound 7:30
55. Brook RD, Franklin B, Cascio W et al (2004) Air pollution and cardiovascular disease – a statement for healthcare professionals from the expert panel on population and prevention science of the american heart association [J]. Circulation 109(21):2655–2671
56. Mar TF, Norris GA, Koenig JQ et al (2000) Associations between air pollution and mortality in phoenix, 1995–1997 [J]. Environ Health Perspect 108(4):347–353
57. Wong TW, Lau TS, Yu TS et al (1999) Air pollution and hospital admissions for respiratory and cardiovascular diseases in Hong Kong [J]. Occup Environ Med 56(10):679–683
58. Takano H, Yanagisawa R, Inoue K et al (2004) Nitrogen dioxide air pollution near ambient levels is an atherogenic risk primarily in obese subjects: a brief communication [J]. Exp Biol Med 229(4):361–364
59. Chen SS, Tang CS, Jin HF et al (2011) Sulfur dioxide acts as a novel endogenous gaseous signaling molecule in the cardiovascular system [J]. Chin Med J 124(12):1901–1905
60. Sunyer J, Ballester F, Le Tertre A et al (2003) The association of daily sulfur dioxide air pollution levels with hospital admissions for cardiovascular diseases in europe (the aphea-ii study) [J]. Eur Heart J 24(8):752–760
61. Peters A, Doring A, Wichmann HE et al (1997) Increased plasma viscosity during an air pollution episode: a link to mortality? [J]. Lancet 349(9065):1582–1587
62. Ayres S M, Giannelli S, Mueller H. Myocardial and systemic responses to carboxyhemoglobin [J]. *Annals of the New York Academy of Sciences*. 1970, 174(1): 268–+
63. Naeher LP, Brauer M, Lipsett M et al (2007) Woodsmoke health effects: a review [J]. Inhal Toxicol 19(1):67–106

64. Boskabady M, Marefati N, Farkhondeh T et al (2018) The effect of environmental lead exposure on human health and the contribution of inflammatory mechanisms, a review [J]. Environ Int 120:404–420
65. Guariguata L, Whiting DR, Hambleton I et al (2014) Global estimates of diabetes prevalence for 2013 and projections for 2035 [J]. Diabetes Res Clin Pract 103(2):137–149
66. Brook RD, Jerreft M, Brook JR et al (2008) The relationship between diabetes mellitus and traffic-related air pollution [J]. J Occup Environ Med 50(1):32–38
67. Kramer U, Herder C, Sugiri D et al (2010) Traffic-related air pollution and incident type 2 diabetes: results from the salia cohort study [J]. Environ Health Perspect 118(9):1273–1279
68. Pearson JF, Bachireddy C, Shyamprasad S et al (2010) Association between fine particulate matter and diabetes prevalence in the u.S [J]. Diabetes Care 33(10):2196–2201
69. Dijkema MB, Mallant SF, Gehring U et al (2011) Long-term exposure to traffic-related air pollution and type 2 diabetes prevalence in a cross-sectional screening-study in the Netherlands [J]. Environ Health Global Access Sci Sour 10:76
70. Coogan PF, White LF, Jerrett M et al (2012) Air pollution and incidence of hypertension and diabetes mellitus in black women living in los Angeles [J]. Circulation 125(6):767–772
71. Andersen ZJ, Raaschou-Nielsen O, Ketzel M et al (2012) Diabetes incidence and long-term exposure to air pollution: a cohort study [J]. Diabetes Care 35(1):92–98
72. Brook RD, Cakmak S, Turner MC et al (2013) Long-term fine particulate matter exposure and mortality from diabetes in Canada [J]. Diabetes Care 36(10):3313–3320
73. Dales RE, Cakmak S, Vidal CB et al (2012) Air pollution and hospitalization for acute complications of diabetes in Chile [J]. Environ Int 46:1–5
74. Eze IC, Schaffner E, Fischer E et al (2014) Long-term air pollution exposure and diabetes in a population-based swiss cohort [J]. Environ Int 70:95–105
75. Zanobetti A, Luttmann-Gibson H, Horton ES et al (2014) Brachial artery responses to ambient pollution, temperature, and humidity in people with type 2 diabetes: a repeated-measures study [J]. Environ Health Perspect 122(3):242–248
76. Robledo CA, Mendola P, Yeung E et al (2015) Preconception and early pregnancy air pollution exposures and risk of gestational diabetes mellitus [J]. Environ Res 137:316–322
77. Malmqvist E, Jakobsson K, Tinnerberg H et al (2013) Gestational diabetes and preeclampsia in association with air pollution at levels below current air quality guidelines [J]. Environ Health Perspect 121(4):488–493
78. Lin YT, Jung CR, Lee YL et al (2015) Associations between ozone and preterm birth in women who develop gestational diabetes [J]. Am J Epidemiol 181(4):280–287
79. Tamayo T, Rathmann W, Kramer U, et al. (2014) Is particle pollution in outdoor air associated with metabolic control in type 2 diabetes? [J]. PloS One 9(3):e91639
80. Chuang KJ, Yan YH, Chiu SY et al (2011) Long-term air pollution exposure and risk factors for cardiovascular diseases among the elderly in Taiwan [J]. Occup Environ Med 68(1):64–68
81. Kim JH, Hong YC (2012) Gstm1, gstt1, and gstp1 polymorphisms and associations between air pollutants and markers of insulin resistance in elderly Koreans [J]. Environ Health Perspect 120(10):1378–1384
82. Thiering E, Cyrys J, Kratzsch J et al (2013) Long-term exposure to traffic-related air pollution and insulin resistance in children: results from the giniplus and lisaplus birth cohorts [J]. Diabetologia 56(8):1696–1704
83. Collaboration N C D R F (2016) Trends in adult body-mass index in 200 countries from 1975 to 2014: a pooled analysis of 1698 population-based measurement studies with 19.2 million participants [J]. Lancet 387(10026):1377–1396
84. Xu Z, Xu X, Zhong M et al (2011) Ambient particulate air pollution induces oxidative stress and alterations of mitochondria and gene expression in brown and white adipose tissues [J]. Part Fibre Toxicol 8:20
85. Liu C, Xu X, Bai Y et al (2014) Air pollution-mediated susceptibility to inflammation and insulin resistance: influence of ccr2 pathways in mice [J]. Environ Health Perspect 122(1):17–26

86. An R, Zhang S, Ji M et al (2018) Impact of ambient air pollution on physical activity among adults: a systematic review and meta-analysis [J]. Perspect Public Health 138(2):111–121
87. Roberts JD, Voss JD, Knight B (2014) The association of ambient air pollution and physical inactivity in the United States [J]. PLoS One 9(3):e90143
88. Montane J, Cadavez L, Novials A (2014) Stress and the inflammatory process: a major cause of pancreatic cell death in type 2 diabetes [J]. Diab Metab Syndrome Obes 7:25–34
89. Kodavanti UP (2015) Air pollution and insulin resistance: do all roads lead to Rome? [J]. Diabetes 64(3):712–714
90. Sun Q, Yue P, Deiuliis JA et al (2009) Ambient air pollution exaggerates adipose inflammation and insulin resistance in a mouse model of diet-induced obesity [J]. Circulation 119(4):538–546
91. Liu C, Fonken LK, Wang A et al (2014) Central ikkbeta inhibition prevents air pollution mediated peripheral inflammation and exaggeration of type ii diabetes [J]. Part Fibre Toxicol 11:53
92. Sun Q, Zhang G, Chen R et al (2018) Central ikk2 inhibition ameliorates air pollution-mediated hepatic glucose and lipid metabolism dysfunction in mice with type ii diabetes [J]. Toxicol Sci 164(1):240–249

Plasticizer Exposure and Reproductive Health: Phthalates and Bisphenol A

Pengpeng Wang and Yan Zhao

Abstract Phthalates and bisphenol A are among the most popular plasticizers used today, which are ubiquitous environmental chemical pollutants with endocrine disruption. In this chapter, we summarize the basic characteristics of phthalates and bisphenol A and their effects on male and female reproductive health. We focus on the effects of phthalates exposure on testosterone level, anogenital distance, semen quality and hypospadias incidence in the male, as well as on precocious puberty, endometriosis, abnormalities of pregnancy in the female. Moreover, the effects of bisphenol A exposure on male semen quality, reproductive cells, sex hormones and female steroid hormone levels, reproductive organ diseases, and adverse birth outcomes are discussed. Results indicate that exposure to phthalates and bisphenol can adversely affect male and female reproductive health. However, evidence is still controversial. More large-scale prospective cohort studies are needed to verify the effects of plasticizer exposure on reproductive health in humans.

Keywords Plasticizer · Phthalates · Bisphenol A · Reproductive health

1 Introduction

Plasticizers are a category of additives that increase the plasticity or decrease the viscosity of materials. These are the substances which are added in order to alter their physical properties, such as durability, elasticity, and flexibility [1]. Phthalates

P. Wang
School of Public Health, Fudan University, Shanghai, China

Key Laboratory of Public Health Safety, Ministry of Education, Fudan University, Shanghai, China

Y. Zhao (✉)
Shanghai First Maternity and Infant Hospital, Tongji University School of Medicine, Shanghai, China

Y. Zhang (ed.), *Emerging Chemicals and Human Health*,
https://doi.org/10.1007/978-981-32-9535-3_4

and bisphenol A (BPA) are among the most popular plasticizers used today, which are widely used in consumer and industrial products. Phthalates and BPA are ubiquitous environmental chemical pollutants with endocrine disruption and adversely affect human health through different mechanisms of action [2], especially potential reproductive toxicity for males and females [3, 4]. This chapter examines the basic characteristics of phthalates and BPA and their effects on human reproductive health.

2 Characterization of Phthalates

Phthalates, the diesters of 1,2-benzenedicarboxylic acid (phthalic acid esters), are a class of synthetic chemicals used as plasticizers and softeners widely in consumer and industrial products, mainly polyvinyl chloride (PVC) or vinyl [5]. More than 25 phthalates are used in commercial applications, with each adding unique qualities to the product into which it is incorporated. Ten commonly used phthalates and their metabolites are presented in Table 1. Phthalates can be classified into two categories

Table 1 Phthalates and their metabolites

Parent compounds (acronym)	Major metabolites	Acronym
Dimethyl phthalate (DMP)	Mono-methyl phthalate	MMP
Diethyl phthalate (DEP)	Mono-ethyl phthalate	MEP
Dibutyl phthalate (DBP)	Mono-n-butyl phthalate	MBP
Benzylbutyl phthalate (BzBP)	Mono-benzyl phthalate	MBzP
Dicyclohexyl phthalate (DCHP)	Mono-cyclohexyl phthalate	MCHP
Di-2-ethylhexyl phthalate (DEHP)	Mono-2-ethylhexyl phthalate	MEHP
	Mono-(2-ethyl-5-hydroxyhexyl) phthalate	MEHHP
	Mono-(2-ethyl-5-oxohexyl) phthalate	MEOHP
	Mono-(2-ethyl-5-carboxypentyl) phthalate	MECPP
	Mono-(2-carboxymethyl-hexyl) phthalate	MCMHP
Diisobutyl phthalate (DiBP)	Mono-isobutyl phthalate	MiBP
Diisononyl phthalate (DiNP)	Mono-isononyl phthalate	MiNP
	Mono-(carboxyisooctyl) phthalate	MCiOP
Diisodecyl phthalate (DiDP)	Mono-(carboxynonyl) phthalate	MCNP
	Mono-(carboxyisononyl) phthalate	MCiNP
Di-n-octyl phthalate (DnOP)	Mono-n-octyl phthalate	MnOP
	Mono-(3-carboxypropyl) phthalate	MCPP
	Mono-carboxy-n-heptyl phthalate	MCHpP
	Mono-n-heptyl phthalate	MHpP
	Mono-n-pentyl phthalate	MPeP
	Mono-iso-propyl phthalate	MiPrP

according to their molecular weight and usage. For example, phthalates with high relative molecular weight include those with 9–13 carbon atoms in their chemical backbone, which gives them increased permanency and durability, such as di(2-ethylhexyl) phthalate (DEHP), diisononyl phthalate (DiNP), and benzylbutyl phthalate (BzBP), and they are mainly used in the manufacture of PVC plastics for food packaging, building materials, and medical devices [6, 7]. Phthalates with low relative molecular weight include those with 3–8 carbon atoms in their chemical backbone, such as diethyl phthalate (DEP) and di-n-butyl phthalate (DnBP), and they are commonly used in the manufacture of personal care products (including perfumes, body lotions, cosmetics, shampoos), paints, and adhesives [8, 9]. In the past 5 years, the global phthalates transaction volume has increased by 30%, from $10.39 billion in 2014 to $13.89 billion in 2018, and the production has increased from 5.35 million tons to 6.76 million tons. The increase in phthalates productions and transactions means that more and more phthalates will enter the environment or the human body in various ways.

Human exposure is widespread across the lifespan, and ingestion, inhalation, and dermal contacts are considered the routes of exposure for the general population. When entering into the human body, phthalate diester may first be hydrolyzed to phthalate monoester, oxidized (phase I metabolism) by the catalytic action of cytochrome P450 (CYP) enzyme, and then converted to glucuronide conjugated phthalate (phase II metabolism). Phthalates which have a short biological half-life can be rapidly metabolized to monoester metabolites after exposure and excreted in the urine [10]. As the environmental endocrine disruptor, phthalates can interfere with hormone synthesis, secretion, binding, and metabolism and have antiandrogenic and estrogen-like activities. Phthalates are also a class of harmful substances with teratogenicity, carcinogenicity, mutagenicity, and reproductive toxicity. Evidence from animal experiments and population-based epidemiological studies suggests that exposure to phthalates can affect human reproductive health [11, 12]. Considering the health hazards of phthalate exposure, U.S. Environmental Protection Agency (EPA) has decided to regulate the manufacture, processing, commercial distribution, and/or use of phthalates.

2.1 *Phthalate Exposure and Male Reproductive Health*

Phthalates have been confirmed to have anti-androgenic activity in animal studies, leading to reduced circulating testosterone and male reproductive tract abnormalities [13]. Male reproductive toxicity is one of the most important toxic effects of phthalates, and phthalate exposure may or may not lead to the "phthalate syndrome" or "testicular dysgenesis syndrome" [14], which is characterized by hypospadias, cryptorchidism, undescended testes, reduced anogenital distance (AGD), decreased sperm count and quality, low testosterone levels, sterility, and testicular cancer [15]. Phthalate exposure during different life stages may cause abnormal levels of hormones in male progeny, lead to decreased sperm quality in adulthood, cause direct or indirect damage to the structure and function of various testicular cells and has toxic effects on testis and its internal structure, ultrastructure, hormone secretion, semen and sperm quality, and other genitals. The health outcomes of fetal testosterone concentrations, male AGD, semen quality, and hypospadias incidence were discussed in this part.

2.1.1 Phthalate Exposure and Testosterone

Testosterone is the primary male sex hormone. In male humans, testosterone plays a key role in the development of the male reproductive system and is necessary for normal male fertility. And testosterone promotes secondary sexual characteristics such as increased bone density, muscle strength, the growth of body hair at the same time. Low testosterone levels are associated with abnormal sexual differentiation, decreased fertility, and other adverse conditions. There is a mass of evidence that fetal phthalate exposure is associated with decreased testosterone in fetuses of rats. A meta-analysis performed by the National Academy of Sciences of the United States showed that a decrease in fetal testes testosterone after in utero exposure to DEHP in rats. The findings are inconsistent in humans [15]. Main et al. found that mono-n-butyl phthalate (MBP) contamination of human breast milk was negatively correlated with free testosterone ($r = -0.22$, $P = 0.033$) and other phthalate monoesters showed similar but nonsignificant tendencies [16]. Lin et al. detected the phthalate metabolite concentrations of MBP, monobenzyl phthalate (MBzP), monoethylhexyl phthalate (MEHP), mono-(2-ethyl-5-hydroxyhexyl)phthalate (5OH-MEHP), mono(2-ethyl-5-oxohexyl) phthalate (5oxo-MEHP), monomethyl phthalate (MMP), and monoethyl phthalate (MEP) in maternal urine samples collected in the third trimester and cord sex steroid hormones in 155 pregnant women and their newborns [17]. The results showed that no significant correlation was found between phthalate metabolites for male newborns and each steroid hormone. Overall, since phthalates are detected in different biological samples and the results from different studies are inconsistent; the evidence that phthalate exposure could decrease fetal testis testosterone is considered indeterminate.

2.1.2 Phthalate Exposure and Semen Quality

Male factor infertility leads to approximately 50% of difficulty with conception and a high proportion of cases are unexplained [18]. Sperm quality is an important factor of difficulty in getting pregnant. Studies have shown that the continuous decrease in sperm count and fertility in men over the past decade may be related to the absorption of more and more phthalates [19]. For the outcomes of sperm count and quality, the results from seven studies showed an inverse relationship between decreased semen quality and increased DEHP exposure. The inverse finding above was also supported by the studies that the DEHP exposure increased apoptosis, reactive oxygen species (ROS) generation, and increased sperm aneuploidy, which may be the potential mechanisms of action [20]. Bloom et al. and Axelsson et al. found that sperm concentration and its motility decreased with increasing exposure to MEOHP and MEHP [21, 22]. Hauser and his colleagues found that urinary levels of MEP and MEHP were associated with DNA damage of sperms [23]. Jurewicz et al. investigated the association of phthalate metabolite levels in urine with semen parameters (sperm concentration, motility, morphology, CASA parameters), sperm aneuploidy, and sperm chromatin structure, and the results demonstrated that urinary phthalate metabolite levels were significantly associated with a decrease in sperm motility (MEHP and MINP), CASA parameters (MBP), testosterone level (MEHP), and an increase in sperm DNA damage (MBP) and sperm aneuploidy (MBzP, MBP, MEHP, and MEP) [24]. Overall, there is robust evidence to confirm the association between increased phthalate exposure and decreased semen quality, specifically sperm concentration.

2.1.3 Effects of Phthalate Exposure on AGD and Hypospadias

AGD is the distance from the midpoint of the anus to the genitalia, the underside of the scrotum or the vagina, which is considered as a significant marker of androgen activity during the male programming window in both humans and animals [25]. The association between phthalate exposure and reduced AGD has been observed in both animals and humans. Of all the studies on phthalate exposure, DEHP exposure is the most studied. The National Academy of Sciences of the United States carried out a systematic review on phthalate exposure and male reproductive tract development in animals from a mass of literature in 2017 and the meta-analyses, meta-regression, and benchmark dose (BMD) estimation were performed in the study. Their research showed that there was consistent evidence for a decrease in AGD in male rats after fetal exposure to DEHP. The BMD analysis showed that the BMD_5 was around 300 mg/kg/day for Sprague Dawley rats, 150 mg/kg/day for Wistar rats, and 250–350 mg/kg/day for mice, suggesting that the Wistar rat was most sensitive to phthalate exposure [26].

As for the effects of phthalate exposure on human reproductive health, prospective cohort studies can provide more convincing evidence. Therefore, here we are only concerned with prospective cohort studies related to phthalate exposure and AGD. In the United States, Swan et al. examined eight phthalate metabolites (DEHP, MEP, MBzP, MBP, MiBP, MCPP, MCNP, and MCOP) in first-trimester maternal urine samples of 753 pregnant women recruited between August 2010 and August 2012 at university-based prenatal clinics in San Francisco, California. Their research presented that the concentrations of DEHP metabolites in first-trimester maternal urine samples were inversely associated with the AGD in male newborns, but not female [27]. Martino-Andrade et al. detected phthalate metabolite concentrations in maternal urine samples collected in three trimesters (the first, second, and third trimesters) in a subset of The Infant Development and the Environment Study mothers ($N = 168$), and the results showed that the concentrations of DEHP metabolites in first-trimester urine samples were inversely related to male AGD [28]. Bustamante-

Montes et al. reported similar findings in Mexico [29]. Radke and his colleagues through the systematic review of human epidemiological evidence showed that there was moderate evidence of an inverse association between AGD and exposure to DEHP and DBP, with mostly consistent results reported among five studies. Evidence for DINP, BBP, DIBP, and DEP was slight [20]. However, Swan et al. showed that MEP, MBP, MBzP, and MiBP were found inversely related to reduced AGD, and no correlation with the AGD was presented in DEHP [30]. In Scandinavia, Jensen et al. found no significant trends toward shorter AGD in boys with higher phthalate exposures, which were measured in the first trimester in this low exposed Danish population [31]. Thus, there is consistent evidence of the shortening of ADG being associated with increased urinary concentrations of phthalate metabolites, especially DEHP.

Hypospadias is a birth defect in boys, resulting in the location of the opening of the penis on the underside instead of at the tip. It is estimated that about 5 boys out of every 1000 born in the United States have hypospadias each year [32]. The overall prevalence of hypospadias was 5.8 boys out of 10,000 male births and presented an increasing trend during 1993–2005 in China [33]. For the relationship between phthalate exposure and hypospadias, Li et al. demonstrated that exposure to DEHP in pregnant rats could cause hypospadias in male rats, and the probability and extent of hypospadias were positively correlated with the exposure dose [34]. However, Jensen et al. conducted a case–control study that was nested within a large biobank of amniotic fluid samples collected in Denmark during 1980–1996 [35]. Their study results showed that there was no significant association between mono(2-ethyl-5-carboxypentyl) phthalate (5cx-MEPP, DEHP metabolite) and hypospadias nor the mono(4-methyl-7-carboxyheptyl) phthalate (7cx-MMeHP, DiNP metabolite) [35]. Chevrier and his colleagues also found that there was no association between hypospadias and urinary concentrations of DEHP metabolites by a nested case-control study [36]. However, these two studies were very small in sample capacity, which limited their power. Overall, evidence from the epidemiological studies that estimate the association between phthalate exposure and hypospadias in male births is limited and inconclusive.

2.2 Phthalate Exposure and Female Reproductive Health

Higher levels of phthalates have been found in women than in men, which revealed an aggregated phthalate exposure in females. General biomonitoring and risk assessment studies in women show that phthalate metabolites can be detected from almost all body fluids in women, such as peripheral blood, urine, saliva, amniotic fluid, follicular fluid, and breast milk [4]. Growing studies have indicated that phthalate exposure is associated with some endocrine-related reproductive outcomes in females during early pregnancy, such as time-to-pregnancy, embryonic loss, pregnancy loss, recurrent pregnancy loss, and thyroid dysregulation. In addition, some related studies have also found that exposure to phthalates is associated with precocious puberty ovarian function, precocious puberty in girls, endometriosis, and female genital tumors [37].

2.2.1 Phthalate Exposure and Precocious Puberty

Precocious puberty is when a child's body begins changing into that of an adult (puberty) too soon. When puberty begins before the age of 8 in girls, it is considered precocious puberty. Puberty means rapid growth of bones and muscles, changes in body shape and size, and development of the body's ability to reproduce. Precocious puberty signs and symptoms in girls include breast growth and first-time menstruation, pubic or underarm hair, rapid growth, adult body odor, and sometimes acne.

Phthalate exposure has been estimated for the association with precocious puberty in many studies. However, the results are inconsistent. Frederiksen et al. detected 12 primary and secondary metabolites of five phthalate diesters in the urine of 725 healthy girls and 25 girls with precocious puberty by one large prospective cohort study to explore the association between phthalate exposure and precocious puberty. Their research presented that the age at pubarche increased with higher phthalate metabolite quartiles (except for MEP) and no significant association between phthalate exposure and breast development was observed [38]. Hashemipour et al. assessed seven phthalate ester metabolites in plasma in 87 girls with precocious puberty and 63 age- and sex-matched controls and the results showed that DEHP metabolite levels were significantly higher in those with precocious puberty than the levels in controls ($P < 0.05$), but this difference was not significant for other phthalate metabolites. However, phthalates detected in plasma may be contaminated by blood-collecting vessels and the results are more doubtful [39]. Thus, it may be difficult to obtain a definite association between phthalate exposure and precocious puberty, based on existing studies.

2.2.2 Phthalate Exposure and Endometriosis

Endometriosis is an estrogen-dependent disease which is characterized by the growth of endometrial epithelial and stromal cells anywhere in the body outside the uterine cavity [40]. Researchers are interested in the potential relationship between phthalate exposure and endometriosis and the role of estrogens in the pathophysiology of endometriosis. Studies have shown that phthalate exposure can reduce endometrial receptivity and implantation sites. In early pregnancy, adequate progesterone and progesterone receptor (PR), downregulation, and relatively low levels of estradiol are necessary for blastocyst implantation. It has been found that DEHP can up-regulate endometrial estrogen receptor alpha (ERα) and progesterone receptor (PR) levels, thereby reducing endometrial receptivity during implantation from animal experiments [41].

Upson et al. investigated urinary phthalate metabolite concentrations and endometriosis risk based on 92 surgically confirmed cases diagnosed between 1996 and

2001 and 195 population-based controls [42]. They detected eight phthalate metabolites (MEHP, MEHHP, MEOHP, MECPP, MBzP, MEP, MiBP, and MnBP) and observed a strong inverse association between urinary MEHP concentration and endometriosis risk (*OR* = 0.3, 95% *CI*: 0.1–0.7). Their results also suggested increased endometriosis risk with greater urinary concentrations of mono-benzyl phthalate (MBzP) and mono-ethyl phthalate (MEP). The studies of Cobellis et al. and Masuyama et al. also found similar results [43, 44]. Kim et al. reported no significant association between DEHP and endometriosis (*OR* = 1.001, 95%*CI:* 1.000–1.002, *P* = 0.161) and a significant weak association between MEHP and endometriosis (*OR* = 1.020, 95%*CI*: 1.003–1.038, *P* = 0.020) by analyzing 97 cases and 169 controls [45]. Reddy et al. reported similar results in 2006 [46]. The studies mentioned above indicate that phthalate exposure could increase the risk of endometriosis. Nevertheless, many other studies have not found the associations between phthalate exposure and endometriosis among study subjects from different countries [47–49]. Therefore, the available literature till now may not support the association between phthalate exposure and endometriosis, and many more studies should be conducted in the future.

2.2.3 Phthalate Exposure and Abnormalities of Pregnancy

Abnormalities of pregnancy such as embryonic loss, spontaneous abortion, recurrent pregnancy loss, and others constitute the most frequent complication of human pregnancy. Growing evidence has indicated that phthalate exposure is associated with some abnormal reproductive outcomes in females.

Wang et al. reported that compared with the control group, the number of implantation embryos in early pregnancy decreased in the MEHP-treated group, so did the number of embryos and live fetuses in middle and late trimesters [50]. Some studies have also found that DEHP causes a significant increase in the abortion rate in the second trimester. Many epidemiological studies also revealed the association between phthalate exposure and abnormalities of pregnancy. Liao et al. recruited 103 patients diagnosed with recurrent pregnancy loss and 76 controls from the Department of Obstetrics and Gynecology in southern Taiwan between August 2013 and August 2017, and 11 phthalate metabolites were analyzed at the same time. The research illustrated that exposure to phthalates could increase the risk of recurrent pregnancy loss with the *OR* = 2.85, *P* = 0.045 [51]. Gao et al. detected seven urinary phthalate metabolites in morning urine during gestational weeks 5–14 from 3220 pregnant women, who were recruited by a prospective cohort study in China. Their research results showed that there were positive associations of MEP, MEOHP, MEHHP, and ΣHMWP with embryonic loss (during gestational weeks 6–10) and only one association of fetal loss (during gestational weeks 11–27) was observed with MEHHP. The findings indicated that Chinese women exposed to phthalates during early pregnancy increased the risk of clinical pregnancy loss, especially embryonic loss [52]. Jukic and his colleagues presented that urinary concentrations of phthalate metabolites were associated with follicular phase length or increased the risk of early pregnancy loss, but there was an association between ΣDEHP and reduced risk of early loss [53]. In summary, most studies support the idea that phthalate exposure can lead to abnormal pregnancy such as spontaneous abortion and recurrent pregnancy loss. It is necessary to reduce phthalate exposure among reproductive-aged women as much as possible.

2.2.4 Phthalate Exposure and Other Reproductive Health Effects

Phthalate exposure is also associated with other aspects of reproductive health or has adverse effects. For example, related studies have confirmed that phthalates can cause the increase of luteinizing hormone and testosterone, inhibit the growth of follicles, and induce immature follicular atresia, leading to ovulation disorder and polycystic ovary morphological changes, suggesting that phthalate exposure may be one of the pathogenic factors of polycystic ovary syndrome [54]. Fu and his colleagues demonstrated that significantly positive associations were observed between DEHP metabolites and risk of breast cancer and uterine leiomyoma, especially for MECPP in breast cancer by standard meta-analysis and bioinformatics analysis [55]. Besides, studies on urinary phthalate metabolite concentrations and reproductive outcomes among women undergoing in vitro fertilization showed that urinary concentrations of DEHP metabolites were inversely associated with oocyte yield, clinical pregnancy, and live birth following assisted reproductive technologies [56]. Therefore, much more health effects of phthalate exposure should be explored in the future, and phthalate exposure should be avoided as much as possible among women during reproductive age.

3 Characterization of BPA

BPA is a general material for the synthesis of plastics, which is widely used in the production of epoxy resin, polycarbonate, and other polymer materials. These polymer materials are widely used in toys, thermal paper, food and beverage packaging, dental protective gel, and other daily necessities or medical supplies. BPA molecules in plastics such as polycarbonates and epoxy resins are released by hydrolysis at high temperatures or under acidic conditions [57]. Combined with the widespread use of plastics in daily life, people are increasingly exposed to BPA. Just like phthalates, BPA also has a short biological half-life and can be rapidly metabolized. The Centers for Disease Control and Prevention (CDC) of the United States and the Canadian Health Measures Survey have reported measurable levels of BPA in urine samples in more than 90% of the participants [58, 59].

Recently, BPA has been widely a cause of concern due to its potential endocrine-disrupting effects and reproductive toxicity. The endocrine-disrupting effects interacting with estrogen receptor α/β (ER α/β) [60], estrogen-related receptor γ, androgen receptor (AR), thyroid hormone receptor, or other various physiological receptors of BPA have been demonstrated by different animal and cell experiments [61]. A mass of epidemiological studies has been performed by researchers all over the world to demonstrate whether BPA has reproductive toxicity. Current results from different animal studies indicate that BPA exposure could decrease sperm production and quality, damage testicular cells, perturb hormone levels, disrupt oocyte quality and maturation, and affect ovary function and uterine morphology [61]. Similar outcomes also have been demonstrated by epidemiological studies [62], and there are also many other health effects caused by BPA. Here, we focus on the effects of BPA exposure on both male and female reproductive health.

3.1 BPA Exposure and Male Reproductive Health

There has been much literature on BPA exposure and male reproductive system. Here we mainly focus on the effects of BPA exposure on male semen quality, reproductive cells, and male sex hormones.

3.1.1 BPA Exposure and Male Sex Hormones

The synthesis of sex hormones is a catalytic reaction of cholesterol as a precursor through a series of steroid hormone synthetases. The main male sex hormones are steroidogenesis including follicle-stimulating hormone (FSH), testosterone (T), estradiol (E2), and luteinizing hormone (LH). BPA can interfere with hormone synthesis.

Till now, there are many epidemiological studies regarding BPA exposure and male sex hormones. Siracusa et al. found the conclusion regarding the relationship between BPA exposure and testosterone in men from a lot of epidemiological studies was inconsistent before 2014, but after that, most studies indicated that BPA exposure could perturb the hormone levels in men [61]. Data from animal experiments suggested that BPA exposure could impair male steroidogenesis in rodents, but the effects varied by species, strain, exposure window, and duration [61]. Some of the results were also demonstrated by some epidemiological studies. For example, Meeker et al. found that increased concentrations of follicle-stimulating hormone and decreased ratios of estradiol to testosterone had associations with high concentrations of BPA in men diagnosed with infertility [63]. However, Mendiola et al. showed significant inverse associations of urinary BPA concentration and free androgen index (FAI) level and the FAI/LH ratio, as well as a significant positive association between BPA and sex hormone-binding globulin [64]. Recently, Liu et al. presented that high BPA exposure was associated with increased prolactin, estradiol, and sex hormone-binding globulin levels in males, potentially contributing to male infertility [65]. Liang et al. proved that men with detectable levels of BPA had a 1.52-fold increased risk (95% *CI*: 1.04–2.21) of having a high LH level

(>75th percentile) compared with men with undetectable levels of BPA, after adjustment for potential confounders [66]. Mustieles et al. found that each natural-log unit increase in urinary BPA concentrations was associated with a 19% increase in geometric mean serum total testosterone levels [67]. Although the results may differ among studies, most of the studies indicate that BPA exposure can disrupt the hormone levels in men.

3.1.2 BPA Exposure and Sperm Quality

Animal experiments show that both prenatal and postnatal BPA exposure decrease sperm quality in mice and rats. Sperm number, sperm apoptosis rate, oxidative stress level, and the alteration of seminiferous tubule morphology are conventional observation markers in previous literature. Recently, genetic markers such as chromosomal abnormalities and meiotic DNA double-strand breaks (DSBs) also have been observed [61]. Compared with the control group, prepubertal male SD rats exposed to BPA presented higher intercellular junction disruptions, sloughing of germ cells, and immature germ cells and cellular debris [68]. In SD rats, BPA exposure at a dose of 50 mg/kg induced higher SOD activity, GSH-Px activity, and MDA level, and BPA exposure at all the other designed doses induced sperm malformation rate and apoptosis in testis [69]. Besides, some researchers found that low-dose BPA exposure induced the accumulation of abnormal chromosomes and meiotic DNA double-strand breaks in the late meiotic stage [70].

Epidemiological data showed that occupationally exposed men from four regions in China and infertility patients recruited from the United States clinics all presented higher urinary BPA levels, which were associated with decreased sperm count and motility. Lassen et al. measured urinary BPA concentrations in 308 young men from the general population and revealed that men in the highest quartile of BPA had significantly lower percentage progressive motile spermatozoa compared with men in the lowest quartile (−6.7%, 95% *CI*: −11.76, −1.63) [71]. Goldstone et al. found that a negative relationship between BPA and DNA fragmentation was the sole significant finding in adjusted linear regression ($\beta = -0.0544$, $P = 0.035$), which suggested less sperm DNA damage. But no evidence that BPA diminishes semen quality was shown in this cohort [72]. Buck Louis et al. found that neither female nor male BPA concentration was associated with time to pregnancy with the OR 0.98(95% CI, 0.86, 1.13) and 1.04 (95% CI, 0.91, 1.18), respectively, by longitudinal investigation of fertility and the environment (LIFE) study [73]. Dodge et al. also showed no significant association between BPA exposure and fertilization or live birth following in vitro fertilization for the first time [74]. Overall, most of the above studies suggest that high exposure to BPA may affect the quality of male sperm, especially decreased sperm motility and forward movement, but no association with other semen quality parameters. However, there is still no strong evidence that BPA exposure can result in infertility.

3.2 BPA Exposure and Female Reproductive Health

The effects of BPA on female reproductive health include ovarian follicle production and quality, female sex steroid hormone levels, ovarian and uterine function and related diseases, abnormal pregnancy outcomes such as time-to-pregnancy, spontaneous abortion, recurrent pregnancy loss endometrium, mammary gland development, and other reproductive tract related diseases. Here, we focus on the impact of BPA on female follicle count, sex steroid hormone levels, and reproductive organ diseases.

3.2.1 BPA Exposure and Ovarian Follicle Number and Quality

Follicle numbers and sex steroid hormone production are required for female fertility.

The effects of BPA exposure on follicular production, maturation, and fertility have been studied by many in vitro and in vivo experiments. BPA reduced ovarian weight and increased the number of multi-oocyte follicles of neonatally exposed lambs, and similar results also were found in macaques [61]. Furthermore, in the in vitro study, Wang et al. showed a decrease in maturation rate when porcine oocytes were exposed to BPA at a dose of 250 μmol/L. [75] Nakano et al. reported that BPA exposure at the dose of 2 μg/mL decreased the maturation rate in ICR mouse oocytes [76]. Besides, results from the Consortium Linking Academic and Regulatory Insights on BPA Toxicity study of the United States showed that BPA exposure at some doses and time points altered follicle numbers and hormone production in female rats [77]. The data mentioned above provide convincing evidence that BPA exposure affects oocyte quality, number, maturation, and induce meiotic arrest of oocytes.

Prospective cohort studies proved that the higher the concentration of BPA in urine, the less the total number of oocytes, mature oocytes, and follicles were normally fertilized, suggesting that BPA can damage the production, maturation and fertilization of oocytes, and reduce the success rate of in vitro fertilization [78, 79].

3.2.2 BPA Exposure and Sex Steroid Hormones

Both epidemiological studies and animal experiments have analyzed associations between BPA exposure and sex steroid hormones such as E2, progesterone(P4), T, and FSH, but the results are not consistent.

The InCHIANTI adult population study showed that BPA exposure was related to changes in estradiol or testosterone levels in women [80]. Lee et al. found urinary BPA levels were associated with higher levels of T, E2, and P4 among individuals with high levels of BPA in Korean girls with precocious puberty. Miao et al. showed that urinary BPA concentrations were positively associated with higher prolactin

(PRL), E2, and P4 levels and a statistically significant inverse association between urinary BPA and FSH in Chinese female workers with occupational BPA exposure [81]. In the United States, BPA was associated with higher total T levels in female adolescents based on the National Health and Nutrition Examination Survey (NHANES) conducted between 2011 and 2012 [82]. Overall, despite the inconsistencies across studies, most of the studies have indicated that BPA levels are associated with alterations of hormone levels in females.

In vivo and in vitro studies confirm that BPA exposure affects sex steroid hormones, including testosterone, progesterone, luteinizing hormone, FHS, and E2. However, its effects vary by animal species, strain, exposure route, and exposure window in vivo, while in vitro studies have indicated that BPA has adverse effects on steroidogenesis [61].

3.2.3 BPA Exposure and Endometriosis

Endometriosis is a gynecological disorder characterized by endometrial glands and stroma that grow outside the uterine cavity. As mentioned above, BPA exposure is associated with E2, indicating that BPA may be involved in the occurrence of estrogen-dependent pathologies. However, the conclusions regarding bisphenol A and endometriosis are not consistent. Cobellis et al. suggested that BPA concentration had an association with the occurrence of endometriosis, but no association between urinary BPA and endometriosis was found by Buck Louis et al. and Itoh et al. [83, 84] Epidemiological studies on BPA and endometriosis are limited and the results are inconsistent, and therefore future studies are needed to elucidate the association between chronic BPA exposure and uterine morphology alterations in the general population.

Polycystic ovary syndrome (PCOS) is a hormonal disorder common among women of reproductive age. Women with PCOS may have infrequent or prolonged menstrual periods or excess male hormone (androgen) levels and can lead to infertility in severe cases. Hu et al. showed that serum BPA may be positively associated with women with PCOS and BPA might be involved in the insulin-resistance and hyperandrogenism of PCOS by a systematic review and meta-analysis [85]. Kandaraki et al. showed higher BPA levels in PCOS women compared to controls and a statistically significant positive association between androgens and BPA, suggesting a potential role of this endocrine disruptor in PCOS pathophysiology [86].

Studies have provided compelling evidence indicating that BPA can adversely affect the uterine both in vivo and in vitro, though the assessment endpoints vary across studies. Signorile et al. and Retha et al. reported that low-dose BPA exposure before pregnancy could induce endometrial polyps and benign or malignant lesions in the uterus of adult mice and effected the Wölffian duct at the same time [87]. Kendziorski and his colleagues found that low-dose BPA exposure of CD-1 mice increased gland nest density, peri-glandular collagen accumulation, and abnormal endometrial epithelial and stromal functions [88].

There are also many other studies on the association between BPA exposure and reproductive organ diseases such as ovarian cancer, breast cancer, and so on. However, most studies are simply correlation analysis, lacking more convincing prospective cohort studies and in-depth mechanisms.

3.2.4 Prenatal Exposure to BPA and Adverse Birth Outcomes

A growing body of evidence from animal studies and humans confirmed that prenatal BPA exposure may be also associated with adverse birth outcomes such as recurrent miscarriage, preterm birth, and decreased birth weight. Lathi RB et al. showed that maternal conjugated BPA exposure was associated with a higher risk of aneuploid and euploid miscarriage [89]. Similar findings were also observed by a study from China [90]. However, a study from *Environmental Health Perspectives* did not show the statistical associations [91]. The relationship between BPA exposure and miscarriage should be confirmed in much more human studies in the future. Researches about BPA exposure and preterm birth or other outcomes also showed similar uncertainties [92]. All in all, the actual impact of BPA exposure on birth outcomes remains unclear untill the present moment.

4 Conclusions

In summary, a large number of studies have been conducted on the effects of phthalate and BPA exposure on male and female reproductive health. In general, both endocrine disruptors have a certain impact on reproductive health, but some results are still controversial. At present, problems in many epidemiological studies are yet to be solved. First, assessments of phthalate and BPA exposure at a single time point cannot represent the true exposure of individuals. Second, most of the studies explore the effects of individual environmental endocrine disruptors on reproductive health, which do not well demonstrate the causality between exposure and outcome. Lastly, the biological samples used to detect exposure to phthalates and BPA are inconsistent, which lead to inability to compare results among different studies. In the future, it is recommended to carry out large-scale prospective epidemiological cohort studies on the effects of BPA exposure on reproductive health and strictly standardize study design and the selection criteria of study subjects.

References

1. Wypych G (2017) Handbook of plasticizers [M], 3rd edn. ChemTec Publishing, Toronto, p xii, 858 pages

2. Yoon K, Kwack SJ, Kim HS et al (2014) Estrogenic endocrine-disrupting chemicals: molecular mechanisms of actions on putative human diseases [J]. J Toxicol Environ Health B Crit Rev 17(3):127–174
3. Minguez-Alarcon L, Hauser R, Gaskins AJ (2016) Effects of bisphenol a on male and couple reproductive health: a review [J]. Fertil Steril 106(4):864–870
4. Benjamin S, Masai E, Kamimura N et al (2017) Phthalates impact human health: epidemiological evidences and plausible mechanism of action [J]. J Hazard Mater 340:360–383
5. Johns LE, Cooper GS, Galizia A et al (2015) Exposure assessment issues in epidemiology studies of phthalates [J]. Environ Int 85:27–39
6. Hauser R, Calafat A M. (2005) Phthalates and human health [J]. Occupat Environ Med 62(11):806–818
7. Just AC, Adibi JJ, Rundle AG et al (2010) Urinary and air phthalate concentrations and self-reported use of personal care products among minority pregnant women in New York city [J]. J Expos Sci Environ Epidemiol 20(7):625–633
8. Swan S H (2007) Human exposure to phthalates and their health effects [J]. International Seminars on Nuclear War and Planetary Emergencies – 36th Session, pp 230–237
9. Machtinger R, Gaskins AJ, Racowsky C et al (2018) Urinary concentrations of biomarkers of phthalates and phthalate alternatives and ivf outcomes [J]. Environ Int 111:23–31
10. Frederiksen H, Skakkebaek NE, Andersson AM (2007) Metabolism of phthalates in humans [J]. Mol Nutr Food Res 51(7):899–911
11. Messerlian C, Wylie BJ, Minguez-Alarcon L et al (2016) Urinary concentrations of phthalate metabolites and pregnancy loss among women conceiving with medically assisted reproduction [J]. Epidemiology 27(6):879–888
12. Messerlian C, Souter I, Gaskins AJ et al (2016) Urinary phthalate metabolites and ovarian reserve among women seeking infertility care [J]. Hum Reprod 31(1):75–83
13. Johnson KJ, Heger NE, Boekelheide K (2012) Of mice and men (and rats): phthalate-induced fetal testis endocrine disruption is species-dependent [J]. Toxicol Sci 129(2):235–248
14. Swan SH (2008) Environmental phthalate exposure in relation to reproductive outcomes and other health endpoints in humans [J]. Environ Res 108(2):177–184
15. National Academies of Sciences E, Medicine. (2017) Application of systematic review methods in an overall strategy for evaluating low-dose toxicity from endocrine active chemicals [M]. The National Academies Press, Washington, DC, p 180
16. Main KM, Mortensen GK, Kaleva MM et al (2006) Human breast milk contamination with phthalates and alterations of endogenous reproductive hormones in infants three months of age [J]. Environ Health Perspect 114(2):270–276
17. Lin LC, Wang SL, Chang YC et al (2011) Associations between maternal phthalate exposure and cord sex hormones in human infants [J]. Chemosphere 83(8):1192–1199
18. Vander Borght M, Wyns C (2018) Fertility and infertility: definition and epidemiology [J]. Clin Biochem 62:2–10
19. Singh S, Li SSL (2012) Epigenetic effects of environmental chemicals bisphenol a and phthalates [J]. Int J Mol Sci 13(8):10143–10153
20. Radke EG, Braun JM, Meeker JD et al (2019) Phthalate exposure and male reproductive outcomes: a systematic review of the human epidemiological evidence (vol 121, pg 764, 2018) [J]. Environ Int 125:606–607
21. Bloom MS, Whitcomb BW, Chen Z et al (2015) Associations between urinary phthalate concentrations and semen quality parameters in a general population [J]. Hum Reprod 30(11):2645–2657
22. Axelsson J, Rylander L, Rignell-Hydbom A et al (2015) Phthalate exposure and reproductive parameters in young men from the general swedish population [J]. Environ Int 85:54–60
23. Hauser R, Meeker JD, Singh NP et al (2007) DNA damage in human sperm is related to urinary levels of phthalate monoester and oxidative metabolites [J]. Hum Reprod 22(3):688–695

24. Jurewicz J, Radwan M, Sobala W et al (2013) Human urinary phthalate metabolites level and main semen parameters, sperm chromatin structure, sperm aneuploidy and reproductive hormones [J]. Reprod Toxicol 42:232–241
25. Thankamony A, Pasterski V, Ong KK et al (2016) Anogenital distance as a marker of androgen exposure in humans [J]. Andrology 4(4):616–625
26. National Academies of Sciences E, Medicine. Application of systematic review methods in an overall strategy for evaluating low-dose toxicity from endocrine active chemicals[M]. National Academies Press, 2017
27. Swan SH, Sathyanarayana S, Barrett ES et al (2015) First trimester phthalate exposure and anogenital distance in newborns [J]. Hum Reprod 30(4):963–972
28. Martino-Andrade AJ, Liu F, Sathyanarayana S et al (2016) Timing of prenatal phthalate exposure in relation to genital endpoints in male newborns [J]. Andrology 4(4):585–593
29. Bustamante-Montes LP, Hernandez-Valero MA, Flores-Pimentel D et al (2013) Prenatal exposure to phthalates is associated with decreased anogenital distance and penile size in male newborns [J]. J Dev Orig Health Dis 4(4):300–306
30. Swan SH, Main KM, Liu F et al (2005) Decrease in anogenital distance among male infants with prenatal phthalate exposure [J]. Environ Health Perspect 113(8):1056–1061
31. Jensen TK, Frederiksen H, Kyhl HB et al (2016) Prenatal exposure to phthalates and anogenital distance in male infants from a low-exposed danish cohort (2010-2012) [J]. Environ Health Perspect 124(7):1107–1113
32. Mai CT, Isenburg J, Langlois PH et al (2015) Population-based birth defects data in the United States, 2008 to 2012: presentation of state-specific data and descriptive brief on variability of prevalence [J]. Birth Defects Res A-Clin Mol Teratol 103(11):972–993
33. Jin L, Ye R, Zheng J et al (2010) Secular trends of hypospadias prevalence and factors associated with it in Southeast China during 1993-2005 [J]. Birth Def Res Part A Clin Mol Teratol 88(6):458–465
34. Li M, Qiu L, Zhang Y et al (2013) Dose-related effect by maternal exposure to di-(2-ethylhexyl) phthalate plasticizer on inducing hypospadiac male rats [J]. Environ Toxicol Pharmacol 35(1):55–60
35. Jensen MS, Norgaard-Pedersen B, Toft G et al (2012) Phthalates and perfluorooctanesulfonic acid in human amniotic fluid: temporal trends and timing of amniocentesis in pregnancy [J]. Environ Health Perspect 120(6):897–903
36. Chevrier C, Petit C, Philippat C et al (2012) Maternal urinary phthalates and phenols and male genital anomalies [J]. Epidemiology 23(2):353–356
37. Kay VR, Chambers C, Foster WG (2013) Reproductive and developmental effects of phthalate diesters in females [J]. Crit Rev Toxicol 43(3):200–219
38. Frederiksen H, Sorensen K, Mouritsen A et al (2012) High urinary phthalate concentration associated with delayed pubarche in girls [J]. Int J Androl 35(3):216–226
39. Hashemipour M, Kelishadi R, Amin MM et al (2018) Is there any association between phthalate exposure and precocious puberty in girls? [J]. Environ Sci Pollut Res 25(14):13589–13596
40. Vercellini P, Vigano P, Somigliana E et al (2014) Endometriosis: pathogenesis and treatment [J]. Nat Rev Endocrinol 10(5):261–275
41. Li R, Yu C, Gao RF et al (2012) Effects of dehp on endometrial receptivity and embryo implantation in pregnant mice [J]. J Hazard Mater 241:231–240
42. Upson K, Sathyanarayana S, De Roos AJ et al (2013) Phthalates and risk of endometriosis [J]. Environ Res 126:91–97
43. Masuyama H, Hiramatsu Y, Kodama JI et al (2003) Expression and potential roles of pregnane x receptor in endometrial cancer [J]. J Clin Endocrinol Metab 88(9):4446–4454
44. Cobellis L, Latini G, Defelice C et al (2003) High plasma concentrations of di-(2-ethylhexyl)-phthalate in women with endometriosis [J]. Hum Reprod 18(7):1512–1515
45. Kim SH, Chun S, Jang JY et al (2011) Increased plasma levels of phthalate esters in women with advanced-stage endometriosis: a prospective case-control study [J]. Fertil Steril 95(1):357–359

46. Reddy BS, Rozati R, Reddy BV et al (2006) Association of phthalate esters with endometriosis in indian women [J]. BJOG 113(5):515–520
47. Weuve J, Hauser R, Calafat AM et al (2010) Association of exposure to phthalates with endometriosis and uterine leiomyomata: findings from nhanes, 1999-2004 [J]. Environ Health Perspect 118(6):825–832
48. Itoh H, Iwasaki M, Hanaoka T et al (2009) Urinary phthalate monoesters and endometriosis in infertile japanese women [J]. Sci Total Environ 408(1):37–42
49. Huang PC, Tsai EM, Li WF et al (2010) Association between phthalate exposure and glutathione s-transferase m1 polymorphism in adenomyosis, leiomyoma and endometriosis [J]. Hum Reprod 25(4):986–994
50. Xin W, Xinli S, Nan W et al (2011) Effect of monoethylhexyl phthalate on embryo implantation during early pregnancy and fetus development during medium and late pregnancy in the rat [J]. Chin J Clin (Electron edn) 5(16):4741–4744
51. Liao KW, Kuo PL, Huang HB et al (2018) Increased risk of phthalates exposure for recurrent pregnancy loss in reproductive-aged women [J]. Environ Poll (Barking, Essex: 1987) 241:969–977
52. Gao H, Zhang YW, Huang K et al. (2017) Urinary concentrations of phthalate metabolites in early pregnancy associated with clinical pregnancy loss in chinese women [J]. Scientific Rep:7
53. Jukic AM, Calafat AM, Mcconnaughey DR et al (2016) Urinary concentrations of phthalate metabolites and bisphenol a and associations with follicular-phase length, luteal-phase length, fecundability, and early pregnancy loss [J]. Environ Health Perspect 124(3):321–328
54. Lovekamp-Swan T, Davis BJ (2003) Mechanisms of phthalate ester toxicity in the female reproductive system [J]. Environ Health Perspect 111(2):139–145
55. Fu ZQ, Zhao F, Chen KL et al (2017) Association between urinary phthalate metabolites and risk of breast cancer and uterine leiomyoma [J]. Reprod Toxicol 74:134–142
56. Hauser R, Gaskins AJ, Souter I et al (2016) Urinary phthalate metabolite concentrations and reproductive outcomes among women undergoing in vitro fertilization: results from the earth study [J]. Environ Health Perspect 124(6):831–839
57. Welshons WV, Nagel SC, Vom Saal FS (2006) Large effects from small exposures. Iii. Endocrine mechanisms mediating effects of bisphenol a at levels of human exposure [J]. Endocrinology 147(6 Suppl):S56–S69
58. Lakind JS, Naiman DQ (2011) Daily intake of bisphenol a and potential sources of exposure: 2005-2006 national health and nutrition examination survey [J]. J Expo Sci Environ Epidemiol 21(3):272–279
59. Haines DA, Murray J (2012) Human biomonitoring of environmental chemicals--early results of the 2007-2009 Canadian health measures survey for males and females [J]. Int J Hyg Environ Health 215(2):133–137
60. Dong S, Terasaka S, Kiyama R (2011) Bisphenol a induces a rapid activation of erk1/2 through gpr30 in human breast cancer cells [J]. Environ Pollut 159(1):212–218
61. Siracusa JS, Yin L, Measel E et al (2018) Effects of bisphenol a and its analogs on reproductive health: a mini review [J]. Reprod Toxicol 79:96–123
62. Minguez-Alarcon L, Hauser R, Gaskins AJ (2016) Effects of bisphenol a on male and couple reproductive health: a review [J]. Fertil Steril 106(4):864–870
63. Meeker JD, Ehrlich S, Toth TL et al (2010) Semen quality and sperm DNA damage in relation to urinary bisphenol a among men from an infertility clinic [J]. Reprod Toxicol 30(4):532–539
64. Mendiola J, Jorgensen N, Andersson AM et al (2010) Are environmental levels of bisphenol a associated with reproductive function in fertile men? [J]. Environ Health Perspect 118(9):1286–1291
65. Liu XQ, Miao MH, Zhou ZJ et al (2015) Exposure to bisphenol-a and reproductive hormones among male adults [J]. Environ Toxicol Pharmacol 39(2):934–941
66. Liang H, Xu WP, Chen JP et al. (2017) The association between exposure to environmental bisphenol a and gonadotropic hormone levels among men [J]. PloS One 12(1)e0169217

67. Mustieles V, Ocon-Hernandez O, Minguez-Alarcon L et al (2018) Bisphenol a and reproductive hormones and cortisol in peripubertal boys: the inma-Granada cohort [J]. Sci Total Environ 618:1046–1053
68. Gurmeet K, Rosnah I, Normadiah MK et al (2014) Detrimental effects of bisphenol a on development and functions of the male reproductive system in experimental rats [J]. EXCLI J 13:151–160
69. Quan C, Wang C, Duan P et al (2017) Bisphenol a induces autophagy and apoptosis concurrently involving the akt/mtor pathway in testes of pubertal sd rats [J]. Environ Toxicol 32(8):1977–1989
70. Liu C, Duan WX, Zhang L et al (2014) Bisphenol a exposure at an environmentally relevant dose induces meiotic abnormalities in adult male rats [J]. Cell Tissue Res 355(1):223–232
71. Lassen TH, Frederiksen H, Jensen TK et al (2014) Urinary bisphenol a levels in young men: association with reproductive hormones and semen quality [J]. Environ Health Perspect 122(5):478–484
72. Goldstone AE, Chen Z, Perry MJ et al (2015) Urinary bisphenol a and semen quality, the life study [J]. Reprod Toxicol 51:7–13
73. Louis GMB, Sundaram R, Sweeney AM et al (2014) Urinary bisphenol a, phthalates, and couple fecundity: the longitudinal investigation of fertility and the environment (life) study [J]. Fertil Steril 101(5):1359–1366
74. Dodge LE, Williams PL, Williams MA et al (2015) Paternal urinary concentrations of parabens and other phenols in relation to reproductive outcomes among couples from a fertility clinic [J]. Environ Health Perspect 123(7):665–671
75. Wang T, Han J, Duan X et al (2016) The toxic effects and possible mechanisms of bisphenol a on oocyte maturation of porcine in vitro [J]. Oncotarget 7(22):32554–32565
76. Nakano K, Nishio M, Kobayashi N et al (2016) Comparison of the effects of bpa and bpaf on oocyte spindle assembly and polar body release in mice [J]. Zygote 24(2):172–180
77. Patel S, Brehm E, Gao L et al (2017) Bisphenol a exposure, ovarian follicle numbers, and female sex steroid hormone levels: results from a clarity-bpa study [J]. Endocrinology 158(6):1727–1738
78. Mok-Lin E, Ehrlich S, Williams PL et al (2010) Urinary bisphenol a concentrations and ovarian response among women undergoing ivf [J]. Int J Androl 33(2):385–393
79. Ehrlich S, Williams PL, Missmer SA et al (2012) Urinary bisphenol a concentrations and implantation failure among women undergoing in vitro fertilization [J]. Environ Health Perspect 120(7):978–983
80. Galloway T, Cipelli R, Guralnik J et al (2010) Daily bisphenol a excretion and associations with sex hormone concentrations: results from the inchianti adult population study [J]. Environ Health Perspect 118(11):1603–1608
81. Miao MH, Yuan W, Yang F et al (2015) Associations between bisphenol a exposure and reproductive hormones among female workers [J]. Int J Environ Res Public Health 12(10):13240–13250
82. Scinicariello F, Buser MC (2016) Serum testosterone concentrations and urinary bisphenol a, benzophenone-3, triclosan, and paraben levels in male and female children and adolescents: Nhanes 2011-2012 [J]. Environ Health Perspect 124(12):1898–1904
83. Buck Louis GM, Peterson CM, Chen Z et al. (2013) Bisphenol a and phthalates and endometriosis: The endometriosis: Natural history, diagnosis and outcomes study [J]. Fertil Steril 100(1):162–169, e161–162
84. Cobellis L, Colacurci N, Trabucco E et al (2009) Measurement of bisphenol a and bisphenol b levels in human blood sera from healthy and endometriotic women [J]. Biomed Chromatogr 23(11):1186–1190
85. Hu Y, Wen S, Yuan DZ et al (2018) The association between the environmental endocrine disruptor bisphenol a and polycystic ovary syndrome: a systematic review and meta-analysis [J]. Gynecol Endocrinol 34(5):370–377

86. Kandaraki E, Chatzigeorgiou A, Livadas S et al (2011) Endocrine disruptors and polycystic ovary syndrome (pcos): elevated serum levels of bisphenol a in women with pcos [J]. J Clin Endocrinol Metab 96(3):E480–E484
87. Signorile PG, Spugnini EP, Mita L et al (2010) Pre-natal exposure of mice to bisphenol a elicits an endometriosis-like phenotype in female offspring [J]. Gen Comp Endocrinol 168(3):318–325
88. Kendziorski JA, Belcher SM (2015) Strain-specific induction of endometrial periglandular fibrosis in mice exposed during adulthood to the endocrine disrupting chemical bisphenol a [J]. Reprod Toxicol 58:119–130
89. Lathi RB, Liebert CA, Brookfield KF et al (2014) Conjugated bisphenol a in maternal serum in relation to miscarriage risk [J]. Fertil Steril 102(1):123–128
90. Shen Y P, Zheng Y M, Jiang J T, et al. (2015) Higher urinary bisphenol a concentration is associated with unexplained recurrent miscarriage risk: evidence from a case-control study in eastern china [J]. PloS One 10(5):e0127886
91. Jukic AM, Calafat AM, Mcconnaughey DR et al (2016) Urinary concentrations of phthalate metabolites and bisphenol a and associations with follicular-phase length, luteal-phase length, fecundability, and early pregnancy loss [J]. Environ Health Perspect 124(3):321–328
92. Mikolajewska K, Stragierowicz J, Gromadzinska J (2015) Bisphenol A – application, sources of exposure and potential risks in infants, children and pregnant women [J]. Int J Occup Med Environ Health 28(2):209–241

Environmental Exposure and Health Effects of Chemical UV Filters

Junyi Ni, Yingya Zhao, and Yunhui Zhang

Abstract More and more attention has been paid to the health effects of UV filter exposure. UV filters are extensively added to personal care products to prevent sunburn, skin aging, skin cancer, and photodegradation caused by UV filter radiation. As the most widely used UV filter, ethyl-hexyl methoxycinnamate (EHMC) is used by 90% of Chinese population. EHMC is lipophilicity, and has endocrine disruptive activity. Although EHMC is widely detected in the environment, there are few reports on human exposure and health risk assessment of EHMC worldwide. In this chapter, we review the environmental exposure and biological effects of UV filters, introduce a method for the detection of EHMC and other UV filters in human urine samples, and assess the exposure level of UV filters in school-aged children. Results show that UV filters are ubiquitous in the environment, bioaccumulation, and have endocrine disruptive effects in organism. UV filters can enter the human body through dietary and skin contact and increase the health risk. Recently, a LC-QTOF-MS method is developed for the determination of EHMC and its main metabolite 4-methoxycinnamic acid in human urine samples. This method is used to detect the exposure level of six kinds of commonly used UV filters in school-aged children. Nine metabolites of four kinds are detected in urine samples, and six of which are detected in more than 50% of urine samples, including BP-3, BP-2, BP-1, EHMC, 4′-MAP, 4-MCA, and OD-PABA. UV filter levels are higher in overweight children than normal weight children, and the exposure levels in girls are significantly

J. Ni
Shanghai Experimental School, Shanghai, China

Y. Zhao
School of Public Health, Fudan University, Shanghai, China

Key Laboratory of Public Health Safety, Ministry of Education, Fudan University, Shanghai, China

Y. Zhang (✉)
Key Laboratory of Health Technology Assessment, National Health Commission of the People's Republic of China, Fudan University, Shanghai, China

Key Laboratory of Public Health Safety, Ministry of Education, Fudan University, Shanghai, China
e-mail: yhzhang@shmu.edu.cn

Y. Zhang (ed.), *Emerging Chemicals and Human Health*,
https://doi.org/10.1007/978-981-32-9535-3_5

higher than those in boys, suggesting that the health risk of girls exposed to UV filters is relatively high currently.

Keywords UV filter · Ethyl-hexyl methoxycinnamate · LC-QTOF-MS measurement · Children health

1 Introduction

As we all know, UV radiation is harmful to the skin. It can hurt the epidermal cells, accelerate the synthesis of pigments, destroy the skin's moisturizing function, dry the skin, damage the elastic fibers in the dermis, and produce fine wrinkles. Under intense irradiation, it can also cause skin inflammation, burns, and even skin cancer. UV filter, an effective component of sunscreen, is a kind of chemical substance that absorbs or reflects UV light to reduce the adverse effects of UV exposure to skin and hair. It is widely used in personal care products and effectively prevents sunscreen, skin aging, and even skin cancer. As a stabilizer, it is also added to cosmetics, plastics, adhesives, rubber, and other products to prevent photodegradation and aging of products. The global annual production of UV filters is up to 10,000 tons, and the consumption is increasing year by year [1]. UV filters can enter aquatic ecosystems directly (through skin erosion) or indirectly (through incomplete treatment of sewage plants). Their continuous, extensive use and potential ecological threats have attracted great attention to environmental scientists and general population recently.

In China, more and more studies on environmental monitoring of UV filters have been carried out in key areas. The monitoring data showed that UV filters were commonly found in different water bodies, such as rivers and lakes, swimming pool water, seawater, groundwater, and drinking water. The concentration of UV filters was as high as 0.3 mg/L. They were also found in sediments, oil, and organisms. The bioaccumulative concentration can reach ng/g dry weight [2]. However, there are few studies on the toxicity effects of UV filters and few data on human exposure which is limited by the detection methods. Thus, it is insufficient to make a health-based risk assessment of UV filters in human.

On the basis of reviewing the environmental exposure and biological effects of UV filters, this study selected ethyl-hexyl methoxycinnamate (EHMC) as the target chemical, developed a detection method of EHMC metabolites and other UV filters in human biological samples, and monitored the exposure level of UV filters in adolescents. The results will provide basic database for understanding the exposure status of UV filters in sensitive population, and promote the health-based risk assessment and environmental standard of UV filters in China.

2 Exposure Assessment of UV Filters in Environment

In 2007, a study firstly reported the concentrations of four commonly used UV filters in Tianjin Wastewater Recycling Plant and discussed different removal methods [3]. In 2011, the researchers measured the concentrations of UV filters (benzotriazole and benzophenone) in sludge of Heilongjiang River Basin (104–6370 ng/g in dry weight) [4]. In 2014, the Leung's team reported the concentrations of UV filters in sea areas worldwide. Among them, Hong Kong, Tokyo, Bangkok, New York, and Los Angeles were the most detectable areas, with Hong Kong having the highest concentration of UV filters, up to 6812 μg/L [5].

EHMC is widely detected in environmental samples. The highest concentrations of EHMC and benzophenone-3 (BP-3) in drinking water were 101 and 115 ng/L in Sao Paulo, Brazil, [6] while the highest levels of these two chemicals in tap water were 290 and 260 ng/L in Barcelona, Spain [7]. Environmental monitoring data showed that EHMC was higher than BP-3, up to 10 times, in waters and seafood, sewage treatment plant inflow, and sediment in Hong Kong and in Norwegian sediment [2, 8]. Compared with BP-3, which is widely used in Europe and America, EHMC is more widely used in China. In 2008, Zhang et al. reported that EHMC was used in 91.3% sunscreen products in China, and the standard-exceeding rate was 0.24% [9]. By contrast, the detection rate of BP-3 in urine in the United States was as high as 96.8% [10]. Human exposure and epidemiological studies of BP-3 were extensively carried out in the United States, which directly led to the European Union to increase the maximum allowable level of BP-3 from below 10% to 6%. However, up to now, there are few monitoring data of EHMC in human samples worldwide.

3 Biological Effects of UV Filters

UV filters have significant endocrine disruptive effects on organisms, inducing vitellogenin, altering the structure of reproductive organs and causing reproductive defects, disrupting gene expression, and even leading to individual malformation and death [11]. EHMC shows endocrine disruptive activity in both in vivo and in vitro experiments. Ecotoxicological data showed that both BP-3 and EHMC could affect aquatic organisms. Compared with BP-3, EHMC was more toxic to algae, sea urchins, sugar shrimp, *Daphnia magna*, and coral. In addition, EHMC at 245 μg/L significantly increased the plasma concentration of vitellogenin in male blackhead fish, and EHMC at 394 μg/L significantly altered the reproductive organs and affected gene expression of the endocrine system, showing estrogen, antiestrogen/androgen activity [12]. Global gene expression profiles showed that 2.2 μg/L EHMC could affect the expression of 1096 genes in zebrafish. These genes were mainly involved in lipid metabolism and synthesis, estrogen metabolism pathway, vitamin A metabolism, DNA damage and death, and cell growth regulation [13].

Therefore, whether EHMC is absorbed through the skin or enriched in the food chain, its risk to human health could not be ignored.

EHMC is more bioaccumulative and potentially more harmful than BP-3. The alcohol/water partition coefficient of EHMC (log $K_{ow} = 5.8$) is much higher than that of BP-3 (log $K_{ow} = 3.5$). EHMC is more lipophilic and easy to enrich in organisms. Previous studies found that the detection rate and concentration of EHMC in fish and shellfish were higher than BP-3 (75% vs. 32%; 5.5–51.3 vs. 3.1–12.4 ng/g dry weight) [14]. Results also showed that the concentration of EHMC was 10 times higher than that of BP-3 in fish, its cumulative content was as high as 1765 ng/g dry weight in mussels, and its concentration was as high as 80 ng/g fat weight in human milk (78% detection rate) [15]. Therefore, it is urgent to make a health-based risk assessment of EHMC.

In recent years, a series of Kannan's studies reported that BP-3 and its derivatives existed in different human samples, such as urine, blood, and adipose tissue. The authors considered that BP-3 could affect the quality of male sperm, cause female endometriosis and uterine fibroids, reduce conception, and even affect the sex ratio of offspring [16]. In 2017, the European Union reduced the permissible limit of BP-3 by 4%. Meanwhile, human exposure and health-based risk assessments of other UV filters were also carried out. More and more attention is paid to EHMC. On March 20, 2015, EHMC was included in the list of trace pollutants monitored by the European Union (2015/495/EU).

4 Method for the Determination of EHMC and Other UV Filters in Human Urine Samples

Because of the high risk of EHMC and its potential endocrine disruption effects, human exposure assessments of EHMC have been proposed with widespread approval. However, EHMC is not easily detected in urine, the most typical human sample, because of its lipophilicity. Previous data have shown that EHMC is rapidly metabolized into other hydrophilic substances within an hour in rats.

In this study, LC-QTOF-MS method is used to deduce the structure of five to six possible metabolites of EHMC in urine samples. One of them is identified as 4-methoxycinnamic acid, which was 100 times higher than that of EHMC in rat urine. Thus, it's used as an exposure marker to detect the actual exposure level of EHMC in human urine samples.

The pretreatment of urine samples includes enzymatic dissociation and extraction of target chemicals, referring to the published method for determination of benzophenone in human urine and blood samples [17]. In brief, 0.5 mL urine sample was taken, and EHMC-d15 (100 μg/L) and 200 mL 1 M ammonium acetate buffer (pH = 4.5) were added to the sample. Then 50 mL glucuronidase/sulfuric acid (20 μg/mL) and 0.5 mL ultrapure water were added and hydrolyzed overnight at 37 °C. After that, the sample was separated and purified by solid-phase extraction

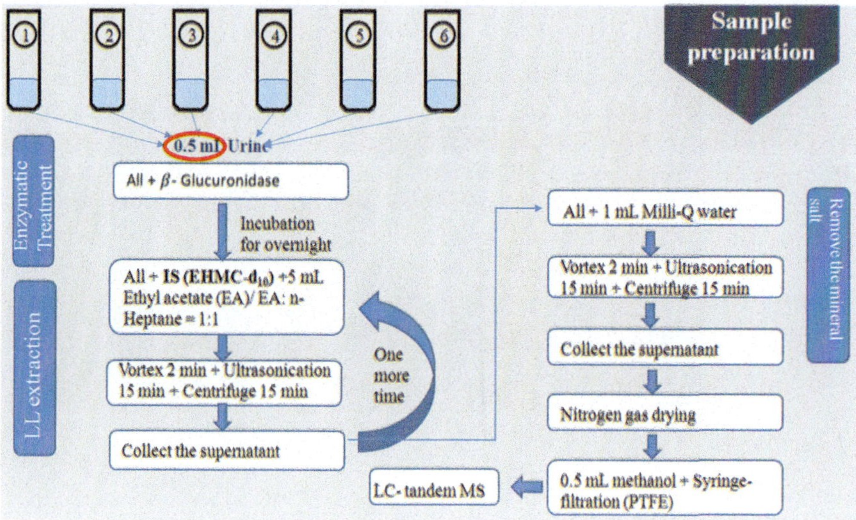

Fig. 1 Flow chart of urine sample pretreatment

column. The target substance was eluted and then dried by nitrogen, and 0.5 mL methanol was added to the eluent to dissolve. The sample was filtered through a 0.2 μm polytetrafluoroethylene (PTFE) filter for instrument determination (Fig. 1).

The advantage of this method is that it can accurately measure the mass-charge ratio of the target chemicals without reference material. With the assistance of ion trap mass spectrometer, the structure of the target chemicals can be deduced by the obtained ion fragments. The polar/nonpolar separation column is selected according to the properties of the target substance. EHMC is a nonpolar substance, so the nonpolar separation column is preferred to separate metabolites of EHMC, and polar separation column is also used for comparison.

The injection volume was 1–5 mL, and the mobile phase was super-pure water (A) and methanol (B) containing 0.1% formic acid. The flow rate was 0.2 mL/min. Gradient elution started at 5% A for 5 min, raised to 50% A in 5 min and kept for 1 min, and then fell back to 5% A in 1 min. The dual spray jet electrospray ionization mode was selected, and the mass ratio of reference to mass ratio was 119.0363 and 966.0070, respectively. The high resolution mass spectrometry (mass-charge ratio, 40–1700) was obtained at 2 MS/s, and tested under positive/negative mode, and the best mode to track the degradation products of EHMC was selected. Nitrogen was used as dry gas and shielding gas. Specific parameters such as flow rate, temperature, and nozzle voltage needed to be further debugged properly by taking characteristics into consideration.

According to the mass-charge ratio of ion trap mass spectrometer, the structure of the metabolites of EHMC was further inferred. The metabolites of EHMC with high ion peak abundance were analyzed and the metabolic pathway map of EHMC was drawn.

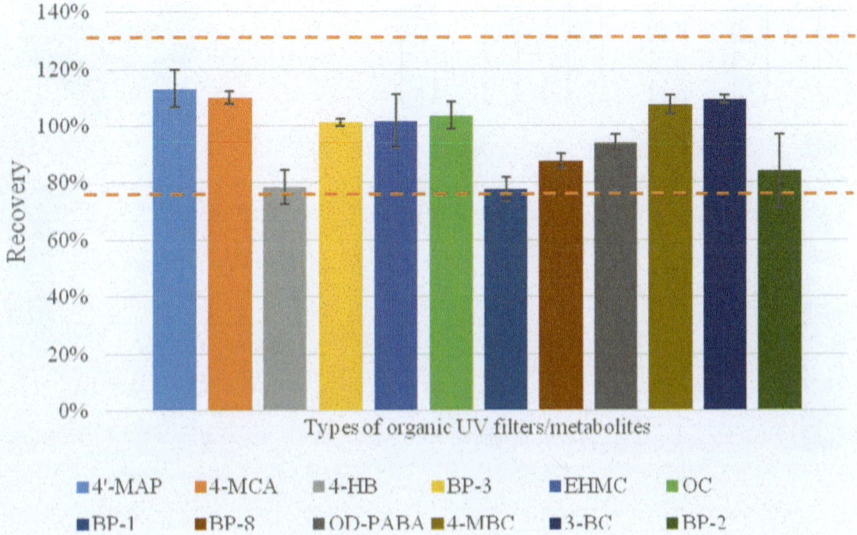

Fig. 2 Recovery rate of UV filter detection in human urine samples

Before instrument analysis, a method was established to correct the parameters, to analyze the feasibility and accuracy of the established experimental method, to calculate the recovery rate, relative standard deviation, instrument/method minimum detection line, and to control the quality of the whole process. The results showed that the recovery rate of urine samples detected by UV filters ranged from 80% to 130% (Fig. 2), suggesting that this method was feasible for the detection of trace UV filters.

5 Exposure Assessment of School-Aged Children Exposed to UV Filters

Due to the limitation of detection technology, only a few European and American research teams detected EHMC in adult biological samples. It was found that the exposure level of EHMC was slightly higher in male adults than that in female adults (0.437 vs. 0.176 ng/mL), and the EHMC concentration in breast milk ranged from 79.85 to 201.00 ng/g. Children and adolescents are more sensitive to environmental chemical exposure than adults. However, data on children's exposure to EHMC are still few.

To assess the internal exposure of UV filters in children, 108 school-aged children aged 8–13 years were enrolled in Jiading District of Shanghai. A questionnaire survey was conducted and 108 urine samples were collected. The samples were stored

Table 1 The detection rates of UV filters in urine samples of school-aged children

	BP-3	4-HB	BP-1	BP-2	BP-8	EHMC	4′-MAP	4-MCA	OD-PABA	4-MBC	3-BC	OC
>LOD (%)	100	0	56	93	7	28	88	88	100	0	0	14
>LOQ (%)	82	0	13	54	0	7	63	35	86	0	0	0

LOD limit of detection, *LOQ* limit of quantification

Table 2 Comparison of the mean concentrations of UV filters in urine samples of school-aged children by different genders

Gender	BP-3	BP-2	BP-1	EHMC	4′-MAP	4-MCA	OD-PABA
Girls	1.37	6.69	2.51	9.47	5.46	8.90	18.50
Boys	1.02	5.48	1.86	3.12	4.13	10.33	10.18

at −20 °C in the dark. Sample providers signed informed consents and filled in questionnaires. The questionnaire included children's age, sex, height, weight, home decoration, parental smoking, breastfeeding, infant pacifier usage, soft plastic toys, family members' perfume usage, children's use of hair gel/perm/hair dye, and nail oil/nail polish, use of toning lotion, materials of microwave oven heating containers, whether plastic containers were often used for water/rice, etc.

Four kinds of UV filters including nine metabolites were detected in urine samples of school-aged children determined by the LC-QTOF-MS method (Table 1). BP-3, BP-2, BP-1, EHMC, 4′-MAP, 4-MCA, and OD-PABA were detected in more than 50% of the urine samples. The exposure levels of UV filters in girls were slightly higher than those in boys (Table 2), suggesting higher exposure level of UV filters and health risk in girls than in boys.

Previous studies showed that the bioaccumulation and toxicity of EHMC was greater than other UV filters. Therefore, we focused on the levels of EHMC and its metabolites 4′-MAP and 4-MCA. Because the exposure data showed non-normal distribution, we assigned 1/2 LOQ to undetected chemicals, and the median value was used in the further stratified comparison. There was a gender difference in the detection rates (Table 3). The detection rates of EHMC and its metabolites in girls were significantly higher than those in boys ($P < 0.05$), but no significant difference in the detection rates of EHMC and its metabolites was found in children of different age and body mass index (BMI) groups. After stratified by gender, concentrations of EHMC and its metabolites were compared among children of different age, BMI, and BMI-Z score (standardized BMI, Z score = [BMI − mean (BMI)]/SD (BMI) groups (Table 4). The levels of EHMC and 4′-MAP detected in overweight girls were significantly higher than those in skinny girls, which was consistent with the lipophilicity of these chemicals.

Table 3 Comparison of detection rates of EHMC and metabolites in urine samples of school-aged children

Classification	4′-MAP			4-MCA			EHMC		
	Number	Detectable rate (%)	*P* value	Number	Detectable rate (%)	*P* value	Number	Detectable rate (%)	*P* value
Age									
≤ 10 years	27	54.0	0.678	16	32.0	0.914	3	6.0	0.415
>10 years	29	50.0		18	31.0		6	10.3	
Gender									
Male	32	56.1	0.346	13	22.8	**0.040**	3	5.3	0.302
Female	24	47.1		21	41.2		6	11.8	
BMI									
Normal	33	55.9	0.408	21	35.6	0.525	4	6.8	0.788
Overweight	12	54.5		5	22.7		2	9.1	
Obesity	11	40.7		8	29.6		3	11.1	
BMI-Z score									
≤0	17	48.6	0.637	12	34.3	0.664	4	11.4	0.42
>0	39	53.4		22	30.1		5	6.8	

P value<0.05

Table 4 Gender-stratified comparison of median levels of EHMC and metabolites in urine samples of school-aged children

Classification	Total children			Boys			Girls		
	4′-MAP	4-MCA	EHMC	4′-MAP	4-MCA	EHMC	4′-MAP	4-MCA	EHMC
Age									
≤10 years	3.030	1.964	0.275	2.976	1.964	0.595	3.838	1.964	0.275
>10 years	2.165	1.964	0.916	3.107	1.964	0.275	1.436	1.964	0.916
BMI									
Normal	3.034	1.964	0.916	3.341	1.964	0.275	1.436	1.964	0.916
Overweight	3.226	1.964	0.595	3.748	1.964	0.595	1.436	1.964	0.595
Obesity	1.436	1.964	0.275	1.436	1.964	0.595	4.400	6.971	0.275
BMI-Z score									
≤0	**1.436**	1.964	**0.275**	3.107	1.964	0.275	**1.436**	1.964	**0.275**
>0	**3.027***	1.964	**0.916***	2.986	1.964	0.275	**3.109***	1.964	**0.916***

* P<0.05, compared with the group of BMI-Z score <=0

6 Conclusions

On the basis of reviewing the environmental exposure and biological effects of UV filters, this chapter introduces a method for the detection of EHMC and other UV filters in human urine samples, and estimates the exposure level of UV filters in school-aged children. UV filters have a wide distribution in the environment, strong bioaccumulation, and significant endocrine disruption effects. These chemicals can enter the human body through dietary and skin contact to increase the health risk of the population. A LC-QTOF-MS method has been developed for the determination of EHMC and its main metabolites in human urine samples. The results show that four kinds of nine metabolites are detected in urine, of which six metabolites have the detection rates of more than 50%, i.e., BP-3, BP-2, BP-1, EHMC, 4′-MAP, 4-MCA, and OD-PABA. UV filters are easily enriched in overweight children, and the exposure levels of EHMC and metabolites in girls are significantly higher than those in boys, suggesting that the health risk of girls exposed to UV filters is relatively high.

The highlight of this study is to systematically assess the environmental and human burden of UV filters, develop a detection method of EHMC and other UV filters in human urine samples, and find out the gender differences of UV filters exposure. These results provide basic data for the prevention of the impact of UV filters on children's health. The study also has limitations, including a small sample size, only focusing on school-aged children and overweight, and a cross-sectional study. Future studies will expand sample size, pay attention to exposure monitoring and health outcomes of sensitive populations, and conduct long-term cohort follow-ups to provide more evidence for the formulation of appropriate environmental standards and prevention strategies.

References

1. Jurado A, Gago-Ferrero P, Vazquez-Sune E et al (2014) Urban groundwater contamination by residues of UV filters. J Hazard Mater 271:141–149
2. Langford KH, Reid MJ, Fjeld E et al (2015) Environmental occurrence and risk of organic UV filters and stabilizers in multiple matrices in Norway. Environ Int 80:1–7
3. Li W, Ma Y, Guo C et al (2007) Occurrence and behavior of four of the most used sunscreen UV filters in a wastewater reclamation plant. Water Res 41:3506–3512
4. Zhang Z, Ren N, Li Y et al (2011) Determination of benzotriazole and benzophenone UV filters in sediment and sewage sludge. Environ Sci Technol 45:3909–3916
5. Tsui MMP, Leung HW, Wai TC et al (2014a) Occurrence, distribution and ecological risk assessment of multiple classes of UV filters in surface waters from different countries. Water Res 67:55–65
6. da Silva CP, Emidio ES, de Marchi MRR et al (2015) The occurrence of UV filters in natural and drinking water in Sao Paulo state (Brazil). Environ Sci Pollut Res 22:19706–19715
7. Diaz-Cruz MS, Gago-Ferrero P, Llora M et al (2012) Analysis of UV filters in tap water and other clean waters in Spain. Anal Bioanal Chem 402:2325–2333
8. Li AJ, Shang Z, Chow CH et al (2017a) Environmental behavior of 12 UV filters and photocatalytic profile of ethy-4-aminobenzoate. J Hazard Mater 337:115–125
9. Wei-Qiang Z, Ying Z, Yu S (2008) Application of sunscreen in sunscreen cosmetics. J Environ Health 25(8):699–701
10. Liao C, Kannan K (2014) Widespread occurrence of benzophenone-type UV light filters in personal care products from China and the United States: an assessment of human exposure. Environ Sci Technol 48:4103–4109
11. Holbech H, Norum U, Korsgaard B et al (2012) The chemical UV-filter 3-benzylidene camphor causes an oestrogenic effect in an in vivo fish assay. Pharmacol Toxicol 91:204–208

12. Christern V, Zucchi S, Fent K (2011) Effects of the UV-filter 2-ethyl-hexyl-4-trimethoxycinnamate (EHMC) on expression of genes involved in hormonal pathways in fathead minnows (Pimephales promelas) and link to vitellogenin induction and histology. Aquat Toxicol 102:167–176
13. Zucchi S, Oggier DM, Fent K (2011) Global gene expression profile induced by the UV-filter 2-ethyl-hexyl-4-trimethoxycinnamate (EHMC) in zebrafish (Danio rerio). Environ Pollut 159:3086–3096
14. Sang Z, Leung KSY (2016) Environmental occurrence and ecological risk assessment of organic UV filters in marine organisms from Hong Kong coastal waters. Sci Total Environ 566-567:489–498
15. Schlumpf M, Kypke K, Wittassek M et al (2010) Exposure patterns of UV filters, fragrances, parabens, phthalates, organochlor pesticides, PBDEs and PCBs in human milk: correlation of UV filters with use of cosmetics. Chemosphere 81:1171–1183
16. Bae J, Kim S, Kannan K et al (2016) Couple's urinary concentrations of benzophenone-type ultraviolet filters and the secondary sex ratio. Sci Total Environ 543:28–36
17. Gao C, Liu L, Ma W et al (2015) Benzonphenone-type UV filters in urine of Chinese young adults: concentration, source and exposure. Environ Pollut 203:1–6

Heavy Metal Exposure and Children's Health

Yuhan Zhou and Wenjuan Ma

Abstract Due to the unique physiological and behavioral characteristics, children are easier to be exposed to environmental pollutants than general population. Children are more sensitive to chemical exposure as their organs are still developing and not yet mature. With the globalization of environmental pollution and the rapid economic growth, the exposure of low concentrations of heavy metals in children is increasingly a global health concern. In this chapter, we systematically review the exposure levels of heavy metals (such as lead, mercury, and cadmium, etc.) and their adverse effects on children's nerve system, kidney, and bone development. As a result, clarifying the major health effects of heavy metals on children will be beneficial for controlling the body burden of heavy metals.

Keywords Heavy metals · Children · Health effects

1 Introduction

The World Health Organization (WHO) has claimed that the global burden of diseases is determined by the patterns of disabilities in low- and middle-income countries (LMICs), which are highly marked with environmental signatures (http://www.who.int/news-room/detail). Heavy metals generally refer to metals with a specific gravity greater than 5 (or a relative density greater than 4.5 g/cm^3), which are widely used in agriculture, national defense, and chemical and pharmaceutical fields. Some heavy metals, such as zinc, manganese, and copper, are essential trace elements for human life, but other heavy metals such as lead, cadmium, and mercury are not

Y. Zhou
School of Public Health, Fudan University, Shanghai, China

Key Laboratory of Public Health Safety, Ministry of Education, Fudan University, Shanghai, China

W. Ma (✉)
MOE-Shanghai Key Laboratory of Children's Environmental Health, Xinhua Hospital, Shanghai Jiao Tong University School of Medicine, Shanghai, China
e-mail: wjma@fudan.edu.cn

Y. Zhang (ed.), *Emerging Chemicals and Human Health*,
https://doi.org/10.1007/978-981-32-9535-3_6

necessary for humans to maintain normal physiological activities, and instead exert certain toxicity to human body. Environmental heavy metal pollution mainly refers to heavy metals such as lead, cadmium, chromium, mercury, and metalloid-like arsenic. Arsenic is a metalloid similar to other heavy metal elements in terms of its nature and toxicity and included as a heavy metal contaminant. Environmental heavy metals mainly come from human activities, such as large-scale disordered mining, smelting and steel plant emission, manufacturing, sewage irrigation, and the use of heavy metal products or metal contaminants, in addition to natural events such as volcanic eruptions and rock weathering. According to a national survey of pollution sources (http://data.stats.gov.cn/easyquery.htm?cn=C01&zb=A0C04&sj=2007), the total amount of chromium, arsenic, lead, cadmium, and mercury in wastewater in 2007 was 25,400 tons, and the emission was 897.3 tons. The emission of five major heavy metal pollutants in the atmosphere—chromium, arsenic, lead, cadmium, and mercury—is about 9500 tons, while the amount of hazardous waste containing lead, mercury, cadmium, chromium, and arsenic is 169,055 million tons. By the end of 2015, the total emissions of five major heavy metal pollutants (lead, mercury, cadmium, chromium and arsenic-like metals) in China was 27.7%, which was lower than in 2007 (http://www.mee.gov.cn/xxgk/hjyw/201611/t20161130_368382.shtml). These heavy metals can enter atmosphere, water and soil, causing serious environmental pollution. For example, data from 387 surface water monitoring stations nationwide, covering 22 provinces (autonomous regions), indicated that heavy metal concentrations in surface water exceeded national standard. In addition, the monitoring data showed that, except lead which was generally a focus in urban air monitoring, other heavy metal contamination in the surrounding atmosphere of some key industrial enterprises was indeed serious. Soil has also been polluted by these heavy metal in different degrees across the country.

With the rapid industrial and economic development, environmental pollution has not been effectively controlled. In addition, the weak awareness of protection from the hazard of heavy metals aggravate heavy metal poisoning incidents. For example, the “itai-itai disease” was caused by cadmium poisoning which happened in Toyama Prefecture, Japan, in 1931 and the Minamata disease was caused by mercury poisoning in Japan in 1956. More than 50,000 cases of heavy metal poisoning happen in China per year. Environmental heavy metal pollution is distinctive from other pollution. Firstly, environmental heavy metals are difficult to pass through the physical, chemical, or biological filtration of nature, reducing or eliminating its toxicity, but have strong enrichment abilities. They are also difficult to degrade in the environment, but can be migrated and enriched into food, vegetables, fruits, and other foods, and then enter into the food chain and further enlarge and enrich its effect. Secondly, the health effects of environmental heavy metals depend not only on their physical and chemical properties, but also on the concentration and valence of the metals. For example, mercury can exert strong neurotoxicity through converting to methylmercury, due to the effect of anaerobic bacteria in the environment [1].

Methylmercury is capable of permanently damaging the central nervous system and posing great health risk to fetuses by passing through blood-brain barrier and placenta [2]. The clinical manifestations of heavy metal poisoning vary according to the type, pathway, dose, and duration of exposure, but more often, the early symptoms are not obvious or lack of differential changes, as a result, missed or delayed diagonosis were caused and risk of disease was then increased.

Children are more susceptible to harmful heavy metals than adults for several reasons. First, children have more ways to contact heavy metals (breastfeeding, placental contact, hand-to-mouth contact, mouth contact). Second, children have higher basal metabolism, a higher food acceptance rate, and a lower toxin elimination rate. Third, since the ventilation per minute is related to the size of the body, children have a higher ventilation per minute than adults, making them easier to inhale more harmful substances (such as secondhand smoke, third-hand smoke, and air pollution; *Pediatric Environmental Health*, the third edition). Fourth, children have larger body surface areas than adults, and as a result, toxic substances can be absorbed through the skin and accumulate in the body. Fifth, children's organs or tissues are developing and therefore are more sensitive to disturbed cellular processes. Sixth, children often have more physical activities to connect with the environment. To sum up, children are vulnerable to heavy metal pollution relative to adults [3]. Heavy metals in the environment are widely present and have a strong accumulation after entering the human body. A large number of studies have shown that the entry of heavy metals into children affects not only children's physical development but also children's nerves, kidneys, and bones. For example, lead can affect almost every system in the body, and even low concentrations of lead can affect the nervous system, cardiovascular system, and kidney system. Both acute and chronic mercury exposure can cause damage to the nervous system, immune system, and kidney [4]. Human body begins to contact and accumulate metal cadmium slowly during childhood, which will result in nephrotoxicity and osteoporosis in adulthood; chronic cadmium exposure can also cause changes in muscle movement and perceived behavioral movement in children [5]. Therefore, it is necessary to study the effects of heavy metal exposure (or combined exposure) on children's health. The health risks of poisonings caused by such five heavy metals as lead, mercury, cadmium, chromium, and arsenic are summarized below.

2 Sources of Heavy Metals

Heavy metal poisons mainly come from the following five aspects. (1) Occupational exposure: heavy metal exposures are common in mining and smelting areas; lead exposure occurs in manufacturing industries such as printing. (2) Environmental pollution: the discharge of industrial waste polluted crops and accumulated in farmed fish. (3) Inhalation of fumes: inhalation of lead from automobile exhaust, inhalation of cadmium from soot, inhalation of various heavy metals and compounds from cigarette smoking, etc. (4) Daily life exposure: cosmetics, hair dyes, herbal preparations containing heavy metals. (5) Exposure by mistake: such as mercury-containing pesticides.

3 Effects of Heavy Metals on Child Growth and Development

The kidney has a large blood supply to quickly enrich toxins in renal cells and tubules, so it is one of the target organs for acute poisoning [6]. Tubulointerstitial or glomerular damage can be induced through direct or indirect mechanisms [7].

For children, the brain is another important target organ for heavy metal poisoning. Because the brain takes a longer time to form than other organs, it is more likely to be affected by environmental pollutants [8]. In addition, during the intrauterine period, the brain, the immune system, and detoxification mechanisms of the fetuses have not been well developed [8–10]. The brain develops rapidly from the second trimester in pregnancy to childhood, during which period the neurons migrate, differentiate, proliferate, and prune; as a result, the neurons are vulnerable to toxic damage. Hence, it is necessary to pay attention to the damage of nerve development during different exposure periods.

Bone metabolism is a continuing normal physiological process and bone turnover is a dynamic equilibrium between bone formation and bone resorption. When bone formation is inhibited or bone resorption is stimulated, the dynamic balance of these two process is disrupted and then lead to osteoporosis. As a metabolically active tissue, bone mass is increased when bone formation rate is greater than bone resorption rate during bone growth and development [11]. Heavy metals can interfere with bone metabolism in children. For example, when cadmium enters the body, it affects the function of kidneys, and then chronic renal insufficiency affects the absorption of calcium, which in turn affects bone growth. In addition, some metals, such as inorganic lead, bind to phosphate, interfere with the binding of calcium ions to phosphates and have an effect on bone growth. Therefore, children's bone development is also susceptible to heavy metals during growth.

3.1 Effect of Lead Exposure on Children's Health

Lead (Pb) is a kind of heavy metal which is nondegradable. It can be accumulated in a long term and can exert toxicity in multi-organs. It is widely used in batteries, ammunition, and radiation protection suits. More importantly, tetraethyl lead has been widely used as a gasoline additive in the world. Since lead is also added to various pigments and color pictures, it is easy for children to contact these items due to their special growth period. Therefore, children are prone to lead exposure with more absorption and less secretion than adults. The main routes of lead poisoning are through respiratory inhalation and digestive tract, such as inhaling lead-containing waste gas, sucking lead-containing paints and toys, taking traditional Chinese medicine preparations with high lead contents, preserved eggs, puffed food, popcorn, canned drinks, etc. A few cases showed that lead could also be absorbed through skin.

Less than 10% of the lead absorbed into blood are stored in plasma while more than 90% are in red blood cells and then are combined to components such as erythrocyte membrane or hemoglobin; after redistribution, 90–95% of lead are stored in the bone. Lead has a long half-life in the bone (5–20 years) and remains relatively stable [12]. The toxic effects of lead are mainly determined by the concentration of lead in the plasma. Lead ions in plasma can bind to plasma proteins and then form diffusing lead-binding protein (Pb-BP), which gradually transfer to the kidney after filtered from glomerulus, accompanied by a certain degree of tubular reabsorption. At last, lead can aggregate in renal tubular epithelial cells. With the increasing concentrations of blood lead, renal tubules can excrete a portion of lead. For organic lead, most of which entered through the digestive tract is excreted through feces, while most of which absorbed into the blood is excreted by the kidney in urine, and a small portion is excreted by the bile. Inorganic lead exerts the main toxic effect on the body, which is a kind of toxicant metabolized in the body. The degradation of lead in the body is extremely slow, and the half-life in the kidney is as long as 531 days [13], so it is easy to cause lead accumulation; in addition, since lead metabolism in the body is similar to calcium, when calcium deficiency or acid-base balance change occurs in the body—the latter is possibly caused by infection, drinking, trauma, taking acidic drugs, and so on—lead phosphate stored in the bone converts to lead hydrogen phosphate, which is 100 times more soluble than lead phosphate, and enters the bloodstream, causing symptoms of lead poisoning and bone diseases (such as osteoporosis).

The renal injury caused by lead mainly manifests as tubulointerstitial damage. (1) Acute injury: short-term, high-dose lead exposure can cause acute injury to proximal tubule function. (2) Chronic injury: long-term, low-dose lead exposure can lead to chronic interstitial nephritis, lead-induced renal hypertension, chronic renal failure, and gout.

Lead poisoning in childhood can interfere with normal bone metabolism and cause functional damage. Lead can reduce the level of active vitamin D in the blood, thereby can interfere with calcium and phosphorus metabolism, and causing a decreasing intestinal absorption and renal excretion of both calcium and phosphorus. With the increase of lead absorption, the level of 1,25-dihydroxy vitamin D_3 [1, 25 $(OH)_2$ D_3] in the blood is lowered. On the one hand, this may be due to insufficient intake of exogenous vitamin D and, on the other hand, lead can damage the 1-alpha-hydroxylation process of vitamin D, which is a necessary step in activating vitamin D, and reduce the activity and quantity of hydroxylase in the liver and kidney. In addition, high levels of lead in blood may promote the decomposition of active vitamin D, thereby lower the concentration of active vitamin D. [14] Some studies have also pointed out that lead may reduce the number of vitamin D receptors on intestinal wall or reduce the sensitivity of active vitamin D. An animal study showed that after feeding a diet containing 0.82% of lead in mice, the level of blood 1,25-$(OH)_2$ D_3 was decreased. A supplement of active vitamin D did no good to the absorption of calcium in small intestine [15]. In addition, lead also directly interferes with calcium-phosphorus metabolism. High levels of lead in the body can cause degeneration of mitochondria in renal cortical proximal tubular epithelial cells and contribute to the loss of bone mineral; what's more, by substituting zinc ions in alkaline phosphatase enzyme, lead can reduce the activity of the enzyme, and thereby affect the deposition of bone salts and result in a decrease in bone mineral content. Lead can also antagonize the action of calcium ions. As a divalent cation, lead has similar biochemical properties to calcium, and the affinity of lead to calcium channels is much greater than that of calcium and calcium channels. Lead ions can eliminate calcium signals by directly replacing calcium ions, interfering with the calcium receptor system, and increasing or decreasing the activity of calcium ion receptor proteins. The close binding of lead to the calcium binding site of osteocalcin inhibits binding of osteocalcin and hydroxyapatite, which can result in the inactivation of osteocalcin, affecting the calcification process of the bone, and ultimately affecting the normal development of the bone [16, 17].

Lead is a neurotoxic substance. Subclinical effects of the central nervous system are the most common health effects. The most profound impact of research is cognitive impairment, which is generally measured by IQ tests. Children's nervous system is especially sensitive to lead. Lead can easily accumulate in their developing brain through the blood-brain barrier and damage the neurophysiological processes. The clinical manifestations of lead poisoning on the developmental nervous system are mainly short attention, inconsistency, and impulsive and aggressive behavior.

Large amounts of literature indicate that lead exposure can impair cognitive function in children. Miranda et al. [18] found that children's blood lead was negatively correlated with the end-of-grade scores in a dose-response manner, with a greater impact on reading performance than mathematical performance. Canfield et al.[14]measured blood lead concentrations and IQ in 172 children. They found that the lifetime average blood lead concentration increasing from 1 to 10 μg/dL was associated with a decrease in intelligence quotient (IQ) of 7.4; when the blood lead concentration was higher than 10 μg/dL, the degree of IQ decline was greater. Lamphear et al. [19] conducted a comprehensive analysis on data collected from seven international population-based longitudinal cohort studies and found that children's blood lead concentrations of 2.4–10, 10–20, and 20–30 μg/dL were associated with IQ declines of 3.9, 1.9, and 1.1, respectively, which was consistent with the previous studies. Rodrigues et al. [20] and Lucchini et al. [21] also found that blood lead was negatively correlated with children's cognitive function. A prospective cohort study in the United States showed that even children whose blood lead levels never exceeded 100 μg/L after birth had an average IQ decrease of 7.4 points at age 5 when their blood levels rose from 10 to 100 μg/L. [14, 19]

Lead has obvious damage on behavior. Boucher et al. [22] have found that lead level in cord blood showed a negative correlation with go and no-go performance of 196 children, suggesting that childhood lead exposure could impair the ability of inhibiting dominant responses. Neugebauer et al. [23] found that maternal blood lead elevation during pregnancy was associated with an increased risk of attention deficit hyperactivity disorder (ADEH) in children. In addition, a longitudinal study including 2195 school-aged children [24] showed that the rise in children's blood lead concentration increased the risk of ADEH. Liu et al. [25] found that children's blood lead was statistically correlated with emotional reaction, anxiety, and pervasive developmental problems.

For prenatal exposure, Sioen et al. [26] measured the concentration of cord blood in a cohort study in Belgium and asked mothers to complete child's Strengths and Difficulties Questionnaire (SDQ) and found that doubling umbilical cord blood lead concentrations were associated with a threefold risk increasing of hyperactivity disorder. Liu et al. [27] assessed 332 mother-child pairs and found that maternal blood lead level in the first trimester was negatively associated with Neonatal Behavioral Neurological Assessment (NBNA) scores at birth.

It is reasonable to deduce that lead exposure has adverse effects on children's cognitive development and behaviors, regardless of prenatal or postnatal exposure. Also, low dose effects should not be ignored. See Table 1.

3.1.1 Effect of Cadmium Exposure on Children's Health

Cadmium (Cd) is usually produced as a by-product of zinc, lead, and copper smelting and is used for batteries, pigments, electroplating, etc. Cadmium can be enriched by aquatic organisms in nature.

Cadmium can enter the human body through food, water, respiratory tract or other ways. Cadmium in the environment can cause "itai-itai disease" by accumulating in the food chain. Cadmium is absorbed into the bloodstream through the digestive tract and respiratory tract. Generally, the digestive tract accounts for 5–10% of cadmium absorption, while the respiratory tract accounts for 15–30% but increases up to 50% in smokers. The absorption rate of cadmium in the digestive tract is affected by the contamination of calcium, iron and protein in food. The half-life of cadmium in the human body counts up to 10–30 years. Cadmium tends to accumulate in the liver and kidney and bonds to metallothionein (MT). Most cadmium is excreted in urine, with a small amount excreted in bile [28, 29]. Animal experiments and population epidemiological investigations have also shown that bone is one of the target organs of chronic toxicity of cadmium and osteoporosis is the main manifestation [30]. Cadmium in the environment can directly or indirectly affect the skeletal system by destroying the dynamic balance of bone formation and resorption.

Cadmium poisoning can inflict kidney damage [20]. Acute renal injury mainly manifests as renal tubular and renal cortical necrosis. Chronic renal injury mainly manifests as chronic interstitial nephritis, and glomerular lesions may occur at advanced stage. After inhaling a massive amount of cadmium dust or cadmium compounds, patients generally experience systemic acute cadmium poisoning such

Table 1 Relationship between lead exposure and children's cognitive function and behavior

References	Research design	Sample size	Country	Biomarkers and age of detection	Interaction	Methods and age	Result
Miranda et al. [18]	Cross-sectional study	86,038,627	The United States	Blood lead, 0–5 years old	No	End-of-grade test, fourth grade	Children's blood lead concentration is negatively correlated with the final test score
Canfield et al. [14]	Prospective cohort study	172,172	The United States	Blood lead, 6, 12, 18, 24, 36, 48, 60 months	No	Stanford-Binet intelligence scale fourth edition, 3, 5 years old	Children's blood lead concentration is negatively correlated with IQ
Lanpher et al. [19]	Combined analysis	15,811,581	The United States, Mexico, Australia, Yugoslavia	Blood lead, 6, 12, 15, 36, 48, 60 months	No	WISC, 4, 7, and 10 years old	Children's blood lead concentration is negatively correlated with IQ
Rodrigues et al. [20]	Prospective cohort study	524,524	The Bangladesh	Blood lead, 20–40 months; water arsenic, water manganese, early pregnancy, 1, 12, 20–40 months	Yes	Bayley-3, 20–40 M Bayley-3, 20–40 M	Children's blood lead is negatively correlated with cognitive level, and the effect of arsenic and lead on cognitive level is found to have a positive multiplication effect.
Lucchini et al. [21]	Cross-sectional study	299,299	The United States	Blood lead, blood manganese, manganese, air manganese, soil manganese, 11–14 years old	Yes	WISC, CASSL, 11–14 years old	Children's blood lead is negatively correlated with IQ planning, and has a weak correlation with the risk of ADHD. No interaction between lead and manganese has been found.
Boucher et al. [22]	Prospective cohort study	196,196	The United States	Cord blood; lead, PCB-153, mercury; blood lead, PCB-153, mercury, 11.3 years old	Yes	GO/NO-GO test, 11.3 years old	High umbilical cord blood lead is positively correlated with go and no-go test error rates; manganese can enhance prenatal lead exposure

Neugebauer et al. [23]	Prospective cohort study	177,177	Germany	PCDD/Fs, PCBs mother: blood lead, PCDD/Fs, PCBs, 32 weeks pregnant; breast milk PCDD/Fs, PCBs	No	KITAP, 8.5 years old	Increased blood lead levels in mothers during pregnancy are associated with an increased risk of ADHD in children
Choi et al. [24]	Prospective cohort study	21,952,195	Korea	Blood lead, 7–9 years old	No	ADHS, 9.5 years old	Children with high blood lead levels have increased risk of ADHD
Liu et al. [25]	Prospective cohort study	13,411,341	China	Blood lead, 3–5 years old	No	CBCL, C-TRF, 6 years old	Blood lead levels are associated with emotional problems, anxiety, and mega-developmental problems
Sioen et al. [26]	Prospective cohort study	270,270	Belgium	Cord blood lead, cadmium, PCBs, dioxins, HCB	No	SDQ, 7–8 years old	Cord blood lead concentration is positively correlated with children's hyperactivity risk

as chemical pneumonia and pulmonary edema. Chronic cadmium poisoning is manifested as hypertension [31], cardiomyopathy, obstructive pulmonary disease, cadmium osteomalacia and/or osteoporosis, abnormal behavior, or decreased intelligence.

Cadmium exhibits neurotoxicity. Obviously, cadmium has adverse effect on cognition. In a cohort study in rural Bangladesh, Kippler et al. [32] assessed the concentration of cadmium in 1305 mothers in early pregnancy and their children at 5 years old. The maternal urinary cadmium concentration was negatively associated with verbal IQ (VIQ), full-scale IQ (FSIQ), and performance IQ (PIQ); however, the associations were different in children urinary samples. In children, the urinary cadmium concentration was negatively associated with VIQ and FSIQ), but not with PIQ. These effects were more pronounced in girls.

As for prenatal exposure, the aforementioned Kippler's [32] study also found that maternal urinary cadmium level was negatively correlated with children's VIQ, PIQ, and FSIQ. In another cohort study conducted in Shandong, China, Wang et al. [33] measured blood cadmium concentration in 300 mothers and the development quotients (DQs) of 149 infants at 12 months of age using Gesell Developmental Schedule. The study found that a tenfold increase in maternal blood cadmium concentrations was associated with a decrease of 5.7 points in infants' social domain DQs. No significant correlations were found between other four domain DQs and maternal blood cadmium levels. Besides, the study also revealed that the level of cadmium in maternal blood was negatively associated with the expression of brain-derived neurotrophic factor (BDNF) in cord serum, which suggested that cadmium may have adverse effects on infants' neurodevelopment.

Cadmium has adverse effects on children's behavior. A cohort study in Belgium [26] revealed that intrauterine exposure to cadmium was associated with abnormal male emotional problem scores. Roberts et al. found that exposure to air cadmium during pregnancy increased the risk of autism spectrum disorder (ASD) in children [34].

It can be seen that both intrauterine and postpartum cadmium exposure have adverse effects on children's cognitive and behavioral abilities. Although studies have speculated that cadmium cannot penetrate the placental barrier [35], the effect of intrauterine cadmium exposure on pediatric neurodevelopment may be act through other pathways. For example, the interfered level of BDNF under intrauterine cadmium exposure [32] suggests potential mechanisms of pediatric neurotoxicity. See Table 2.

The most prominent toxic effects of cadmium are beosteomalacia and osteoporosis. Available studies on cadmium-induced bone damage mainly focus on Japanese patients with "itai-itai disease," manifested as back and leg pain, osteoporosis and pathological fractures, as well as systemic manifestations such as bloating and indigestion. Cadmium causes bone decalcification and decreased bone density by interfering with the metabolism of calcium, vitamin D, and collagen, and finally results in osteoporosis or osteoporosis [36]. Osteoporosis is also related to kidney damage. The cadmium-induced kidney damage can lead to a decrease in the reabsorption of calcium and phosphorus in the kidney and an abnormal metabolism of vitamins,

Table 2 Relationship between cadmium exposure and children's cognitive function and behavior

References	Research design	Sample size	Country	Biomarkers and age of detection	Interaction	Ending measuring machine measuring age	Result
Kippler [32]	Prospective cohort study	1305	Bangladesh	Mother: Urinary cadmium, early pregnancy; child: Urinary cadmium, 5 years old	No	WPSI, SDQ, 5 years old	Urinary cadmium in children and urinary cadmium during pregnancy are negatively correlated with VIQ, PIQ and FSIQ
Wang [33]	Prospective cohort study	149	China	Mother: Blood cadmium, the day of production	No	GDS, 12 months	Mother blood cadmium is negatively correlated with infant society
Sioen [26]	Prospective cohort study	270	Belgium	Cord blood lead, cadmium, PCBs, dioxins, HCB	No	SDQ, 7–8 years old	Intrauterine cadmium exposure is associated with boys' emotional problems
Roberts [34]	Nested case-control study	Case 325, control 22,101	The United States	US EPA environmental monitoring data; air lead, cadmium, manganese, etc.	No	ADI-R	Cadmium exposure during pregnancy increases the risk of ASD in children

which indirectly promotes the osteotoxic effect. After exposure, the adverse effect of cadmium on osteoblasts still exists, which is manifested by the continuous inhibition to osteoblast proliferation, differentiation, and mineralization [37].

3.2 Combined Exposure and Interaction of Lead and Cadmium

Zabran et al. [38] examined various heavy metals in soil and found that the concentration of nickel, cobalt, manganese, copper, cadmium, chromium, zinc, and lead were all negatively associated with children's academic performance. Kim et al. [39, 40] examined maternal blood lead and blood cadmium concentrations in early and late pregnancy in a cohort study and evaluated neurocognitive development of 6-month-old infants using the second edition of the Bayley Infant Development Scale. The results showed that the maternal blood concentrations of lead and cadmium during early pregnancy had antagonism effects on infants' mental development index (MDI), while the maternal metal exposure in late pregnancy had synergistic effects. Animal experiments also showed that the combined effects of lead and cadmium exposure could cause long-term damage o children's neurodevelopment. [41–43]

In addition to lead and cadmium, there are many other elements affecting the neurodevelopment of children, with interactions. For instance, lead and manganese have a synergistic effect on neurodevelopment [44], probably because of the same pathway or cumulative effects. With increasing concentrations of lead, the adverse effects of arsenic on cognitive function aggravate. [19] Selenium can antagonize the toxic effects of many heavy metals [45]. Details are shown in Table 3.

3.3 Effect of Mercury Exposure on Children's Health

Mercury (Hg) poisoning can be categorized into occupational and nonoccupational poisoning. Nonoccupational mercury poisoning is often caused by taking mercury-containing remedies (such as cinnabar, excessive realgar nanoparticles), intaking mercury-contaminated foods, or using mercury-containing cosmetics [46]. Chronic mercury poisoning can occur when infants and young children are exposed to or ingest mercury-containing substances such as paints in houses and insect repellents. Mercury poisoning does not have clinically specific indicators, and as a result, the cases are easily misdiagnosed and missed. Mercury can enter the body through the respiratory tract, skin, and digestive tract. Mercury vapor is fat-soluble and can enter the bloodstream through the alveolar membrane. The half-life of mercury is 70 days. The absorption rate of organic mercury (mainly methylmercury) in the digestive tract is close to 100%, while the absorption rate of inorganic mercury is

Table 3 Relationship between joint exposure and children's cognition and behavior

References	Research design	Sample size	Country	Biomarkers and age of detection	Interaction	Ending measuring machine measuring age	Result
Zahran [38]	Ecological research	119	The United States	Soil: lead, cadmium, chromium, zinc, cobalt, manganese	Yes	School performance, grade 4	Heavy metal exposure in soil was negatively correlated with school performance scores, and 22–24% fractional changes were associated with heavy metal exposure in soil.
Kim [39]	Prospective cohort study	884	Korea	Mother: blood lead, blood cadmium, early pregnancy and late pregnancy	Yes	BSID-2, 6 months	Intrauterine lead and cadmium exposure are negatively correlated with neurodevelopment in children, and there is interaction between exposures.
Henn [44]	Prospective cohort study	455	Mexico	Children's blood lead, blood manganese, 12, 24 months	Yes	BSID-2, 12, 18, 24, 30, 36 months	Lead and manganese have synergistic effects on neurodevelopment

15%. Little metallic mercury is absorbed by the digestive tract. Mercury largely accumulates in the kidney and brain. Inorganic mercury in the body exists in the form of mercury metallothionein, with 80% in the kidney. However, organic mercury can enter the central nervous system through the blood-brain barrier and enter the fetus through the placenta (http://www.uptodate.com/contents/epidemiologyand-toxicity-of-mercury), posing great harm to fetal growth.

Mercury is a highly neurotoxic pollutant. The Minamata disease was happened in the early twentieth century, when residents had long consuming methylmercury-contaminated aquatic products in Minamata Bay, Kumamoto Prefecture, Japan. Methylmercury is lipophilic, cumulative, and highly neurotoxic. It can enter and accumulate in the brain through the blood-brain barrier and transfer into the fetus through the placenta. Epidemiological studies have proven that prenatal exposure to methylmercury leads to offsprings' development retardance, low body weight, cerebral palsy, epilepsy, etc.; childhood exposure to methylmercury leads to delayed neurotoxicity symptoms, such as limb paralysis, movement disorders, gait abnormalities, memory loss, and even severe mental disorders. [47] Since organic mercury is very easy to combine with sulfhydryl due to its fat structure, it does great harm to neurodevelopment. It can trigger oxidative stress in the brain [48], affect neurotransmitter transmission, induce apoptosis of neuronal cells and degenerative changes, and damage the brain growth. Organic mercury is a powerful teratogen, which can destroy the normal migration pattern of nerve cells and the histological structure of nerve cells in the developing brain.

The renal toxicity of mercury can be divided into acute and chronic injuries. (1) Acute injury: short-term, high-dose mercury intake can lead to acute tubular necrosis, manifested as acute interstitial nephritis and acute renal failure. (2) Chronic injury: long-term, low-dose intake caused by bioaccumulation, in addition to renal tubular damage, can lead to mesangial lesions and mesangial proliferative lesions. Tubulointerstitial lesions are shown as oliguric acute renal injury, while glomerular lesions are manifested as proteinuria, nephrotic syndrome, and renal dysfunction.

Other manifestations of mercury poisoning include for nervous system, paresthesia, ataxia, tremor, and even visual hearing loss and coma; for digestive system, stomatitis, and corrosive gastroenteritis; and for respiratory system, respiratory damage.

3.4 *Effect of Chromium Exposure on Children's Health*

Chromium (Cr), a brittle and hard gray metal, can form compounds with different valence. In nature, chromium mainly exists as a trivalent state, and hexavalent chromium is the most toxic. Evidence has shown that hexavalent chromium compounds cannot be degraded naturally but can accumulate in human body for a long time. The compounds of hexavalent chromium are poisonous and can cause cancer and induce gene mutation. The U.S. Environmental Protection Agency (EPA) has identified hexavalent chromium as one of 17 most dangerous carcinogens.

Chromium poisoning can be categorized into occupational and nonoccupational poisonings. The main routes of occupational chromium poisoning are breathing dust and fumes of chrome ore and chrome smelting, and the routes of nonoccupational poisoning are via absorption of chromium contaminated food and water, skin contact, or digestive tract.

Chromium can bond to ferrous globulin or albumin in plasma to form hexavalent chromium. It has a strong oxidative effect and can be easily absorbed and accumulate in the body. Inhaled chromium mainly accumulates in the lung, kidney, and spleen, while absorbed chromium through the digestive tract and skin contact mainly accumulates in the kidney, liver, spleen, and bone. Chromium is excreted in urine, with a small amount in feces.

Chromium poisoning is manifested as kidney damage and other systemic damage. Acute kidney injury is mainly characterized by oliguric acute renal injury with impaired renal tubular function. Other systemic injuries are manifested as skin burns, atopic dermatitis or asthma, jaundice, liver damage, and tumors of respiratory tract. Chronic poisoning is mainly manifested as chrome sores, nasal septal perforation, dermatitis, and eczema. Recently, a cohort study has revealed that chromium may be a toxic metal for fetal growth measured by abdominal circumference and estimated fetal weight, and early and mid-pregnancy seem to be the most vulnerable period for fetal exposure to chromium [49].

3.5 *Effect of Arsenic Exposure on Children's Health*

Arsenic is a constituent element of the earth's crust. It is widely distributed in rocks, soils, and water environments in nature and exists mostly in the form of arsenic-containing ores such as ornithine (arsenic disulfide) and orpiment (arsenic trisulfide). As a kind of metal-like environmental pollutant, arsenic mainly comes from the mining, smelting, and processing of chemical plants containing arsenic ore and related industries. Arsenic is a carcinogen first identified by the International Agency for Research on Cancer in 1987. Arsenic affects multiorgan systems. It has a strong accumulation in the human body, which can cause damage to the liver, kidney, brain, and cardiovascular and other systems. More and more epidemiological surveys have shown that low doses of arsenic exposure are associated with the development of the child's nervous system [50]. Early arsenic exposure in fetuses and children is associated with bronchiectasis in adulthood [51]. A cross-sectional survey of children aged 8–11 years showed that, after controlling mothers' age and IQ, children's school attendance, head circumference, family environment, and other confounding factors, the level of blood arsenic was negatively correlated with verbal intelligence. After creatinine correction, urinary arsenic was negatively associated with comprehensive intelligence, verbal intelligence, and operative intelligence, though the corrections mentioned above were modified and weakened after adjusting the possible confounders [52]. Another study showed that the level of blood arsenic was negatively correlated with overall athletic ability, fine manual control, and physical coordination, [53] and the concentration of hair arsenic was negatively correlated with speech intelligence and general intelligence [54]. Negative associations were also found between urinary arsenic concentration and neurobehavioral problem-solving ability, verbal memory and digital test capability, sorting capability, and visual tracking ability [55]. In a case-control study, the concentrations of blood and urinary arsenic showed no significant differences between autistic children and control ones aged from 5 to 16 years old [56]. However, the level of blood arsenic is negatively associated with common nervous system development in autistic patients. In a study launched in a mining district, the score of intelligence, operational intelligence, ranking, and spatial ability of those arsenic exposed ones were significantly lower than controls aged 6–9 years old. After adjusting possible confounders, the level of urinary arsenic was significantly associated with comprehensive intelligence and verbal intelligence [57]. Environmental exposure studies also showed that children's overall IQ and operational IQ were negatively correlated with arsenic level in water [58, 59].

4 Summary and Perspectives

Since heavy metals naturally exist in the environment and are easy to accumulate in the body from food chain, it is difficult to control their health risks, especially their adverse effects on metabolism and nervous system. Children, as the sensitive population, are widely exposed to heavy metals and their nerve system, kidney function and bone development are impaired by lead, cadmium, mercury, chromium, arsenic, etc. The combined exposure of various heavy metals and its health risk on children is needed to be assessed in the future cohort studies.

However, limitations and challenges are existing in the study field, including:

1. Most of the studies focus on the effect of only one or two heavy metals. Nonetheless, people are exposed to a mixture of heavy metals with other kinds of environmental pollutants; also, the interactions between these pollutants add to the difficulties in evaluating the real effects. Therefore, it is essential to study the combined effects of different pollutants.
2. Although abundant epidemiological and toxicological studies have revealed that heavy metals pose adverse effects on children's health, the long-term and low-dose effects have not been well elucidated, which require new methods for evaluation. In fact, computational toxicology is now expected to open up new horizons for the association between low-dose, long-term, mixed exposure, and health hazards in humans. In addition, due to regional differences, strengthening trans-regional cooperation and establishing a comprehensive exposure and health effect assessment system will help to eliminate the inconsistency among studies and provide strong data support for pollution prevention and control policies.

References

1. Clarkson TW, Magos L (2006) The toxicology of mercury and its chemical compounds [J]. Crit Rev Toxicol 36(8):609–662
2. Mercury study report to congress, executive summary (1997)
3. Tan SY, Praveena SM, Abidin EZ et al (2016) A review of heavy metals in indoor dust and its human health-risk implications [J]. Rev Environ Health 31(4):447–456
4. Bose-O'reilly S, Mccarty KM, Steckling N et al (2010) Mercury exposure and children's health [J]. Curr Probl Pediatr Adolesc Health Care 40(8):186–215
5. Schoeters G, Den Hond E, Zuurbier M et al (2006) Cadmium and children: exposure and health effects [J]. Acta Paediatr Suppl 95(453):50–54
6. Van Vleet TR, Schnellmann RG (2003) Toxic nephropathy: environmental chemicals [J]. Semin Nephrol 23(5):500–508
7. Jomova K, Valko M (2011) Advances in metal-induced oxidative stress and human disease [J]. Toxicology 283(2–3):65–87
8. Rodier PM (1995) Developing brain as a target of toxicity [J]. Environ Health Perspect 103:73–76
9. Grandjean P, Landrigan PJ (2014) Neurobehavioural effects of developmental toxicity [J]. Lancet Neurol 13(3):330–338
10. Selevan SG, Kimmel CA, Mendola P (2000) Identifying critical windows of exposure for children's health [J]. Environ Health Perspect 108(Suppl 3):451–455
11. Lanphear BP (2015) The impact of toxins on the developing brain [J]. Annu Rev Public Health 36:211–230
12. Rojas M, Espinosa C, Seijas D (2002) Toxicological profile of children monitored for lead exposure. Cituc, 1998-2000 [J]. Epidemiology 13(4):S208–S208
13. Nilsson U, Attewell R, Christoffersson JO et al (1991) Kinetics of lead in bone and blood after end of occupational exposure [J]. Pharmacol Toxicol 68(6):477–484
14. Canfield RL, Henderson CR Jr, Cory-Slechta DA et al (2003) Intellectual impairment in children with blood lead concentrations below 10 microg per deciliter [J]. N Engl J Med 348(16):1517–1526
15. Rosen JF, Chesney RW, Hamstra A et al (1981) Reduction in 1,25-dihydroxyvitamin-d in children with increased lead absorption [J]. Gen Pharmacol 12(2):A10–A10
16. Ohsawa M (2009) Heavy metal-induced immunotoxicity and its mechanisms [J]. Yakugaku Zasshi-J Pharmaceut Soc Japan 129(3):305–319
17. Needleman HL (1988) The persistent threat of lead: medical and sociological issues [J]. Curr Probl Pediatr 18(12):697–744
18. Miranda ML, Kim D, Galeano MAO et al (2007) The relationship between early childhood blood lead levels and performance on end-of-grade tests [J]. Environ Health Perspect 115(8):1242–1247
19. Lanphear BP, Hornung R, Khoury J et al (2005) Low-level environmental lead exposure and children's intellectual function: an international pooled analysis [J]. Environ Health Perspect 113(7):894–899
20. Rodrigues EG, Bellinger DC, Valeri L et al (2016) Neurodevelopmental outcomes among 2-to 3-year-old children in Bangladesh with elevated blood lead and exposure to arsenic and manganese in drinking water [J]. Environ Health 15:44
21. Lucchini RG, Zoni S, Guazzetti S et al (2012) Inverse association of intellectual function with very low blood lead but not with manganese exposure in italian adolescents [J]. Environ Res 118:65–71
22. Boucher O, Burden MJ, Muckle G et al (2012) Response inhibition and error monitoring during a visual go/no-go task in inuit children exposed to lead, polychlorinated biphenyls, and methylmercury [J]. Environ Health Perspect 120(4):608–615

23. Neugebauer J, Wittsiepe J, Kasper-Sonnenberg M et al (2015) The influence of low level pre- and perinatal exposure to pcdd/fs, pcbs, and lead on attention performance and attention-related behavior among german school-aged children: results from the Duisburg birth cohort study [J]. Int J Hyg Environ Health 218(1):153–162
24. Choi WJ, Kwon HJ, Lim MH et al (2016) Blood lead, parental marital status and the risk of attention-deficit/hyperactivity disorder in elementary school children: a longitudinal study [J]. Psychiatry Res 236:42–46
25. Liu JH, Liu XC, Wang W et al (2014) Blood lead concentrations and children's behavioral and emotional problems a cohort study [J]. JAMA Pediatr 168(8):737–745
26. Sioen I, Den Hond E, Nelen V et al (2013) Prenatal exposure to environmental contaminants and behavioural problems at age 7-8 years [J]. Environ Int 59:225–231
27. Liu JA, Gao DG, Chen YM et al (2014) Lead exposure at each stage of pregnancy and neurobehavioral development of neonates [J]. Neurotoxicology 44:1–7
28. Prozialeck WC, Edwards JR (2012) Mechanisms of cadmium-induced proximal tubule injury: new insights with implications for biomonitoring and therapeutic interventions [J]. J Pharmacol Exp Ther 343(1):2–12
29. Wilson AK, Bhattacharyya MH (1997) Effects of cadmium on bone: an in vivo model for the early response [J]. Toxicol Appl Pharmacol 145(1):68–73
30. Jarup L, Alfven T, Persson B et al (1998) Cadmium may be a risk factor for osteoporosis [J]. Occup Environ Med 55(7):435–439
31. Uriu K, Morimoto I, Kai K et al (2000) Uncoupling between bone formation and resorption in ovariectomized rats with chronic cadmium exposure [J]. Toxicol Appl Pharmacol 164(3):264–272
32. Kippler M, Tofail F, Hamadani JD et al (2012) Early-life cadmium exposure and child development in 5-year-old girls and boys: a cohort study in rural Bangladesh [J]. Environ Health Perspect 120(10):1462–1468
33. Wang YW, Chen LM, Gao Y et al (2016) Effects of prenatal exposure to cadmium on neurodevelopment of infants in Shandong, China [J]. Environ Pollut 211:67–73
34. Roberts AL, Lyall K, Hart JE et al (2013) Perinatal air pollutant exposures and autism spectrum disorder in the children of nurses' health study ii participants [J]. Environ Health Perspect 121(8):978–984
35. Markle JGM, Frank DN, Mortin-Toth S et al (2013) Sex differences in the gut microbiome drive hormone-dependent regulation of autoimmunity [J]. Science 339(6123):1084–1088
36. Chen Z, Myers R, Wei TY et al (2014) Placental transfer and concentrations of cadmium, mercury, lead, and selenium in mothers, newborns, and young children [J]. J Expos Sci Environ Epidem 24(5):537–544
37. Scott R, Cunningham C, Mclelland A et al (1982) The importance of cadmium as a factor in calcified upper urinary-tract stone disease – a prospective 7-year study [J]. Br J Urol 54(6):584–589
38. Zahran S, Mielke HW, Weiler S et al (2012) Associations between standardized school performance tests and mixtures of pb, zn, cd, ni, mn, cu, cr, co, and v in community soils of New Orleans [J]. Environ Pollut 169:128–135
39. Kim Y, Ha EH, Park H et al (2013) Prenatal lead and cadmium co-exposure and infant neurodevelopment at 6 months of age: the mothers and children's environmental health (moceh) study (vol 35, pg 15, 2013) [J]. Neurotoxicology 37:248–249
40. Cory-Slechta DA, Weiss B, Cranmer J (2008) The environmental etiologies of neurobehavioral deficits and disorders: weaving complex outcomes and risk modifiers into the equation [J]. Neurotoxicology 29(5):759–760
41. Antonio MT, Corpas I, Leret ML (1999) Neurochemical changes in newborn rat's brain after gestational cadmium and lead exposure [J]. Toxicol Lett 104(1–2):1–9
42. Antonio MT, Lopez N, Leret ML (2002) Pb and cd poisoning during development alters cerebellar and striatal function in rats [J]. Toxicology 176(1–2):59–66

43. Leret ML, Millan JA, Antonio MT (2003) Perinatal exposure to lead and cadmium affects anxiety-like behaviour [J]. Toxicology 186(1–2):125–130
44. Henn BC, Schnaas L, Ettinger AS et al (2012) Associations of early childhood manganese and lead coexposure with neurodevelopment [J]. Environ Health Perspect 120(1):126–131
45. Tinggi U (2008) Selenium: its role as antioxidant in human health [J]. Environ Health Prev Med 13(2):102–108
46. Chan TYK (2011) Inorganic mercury poisoning associated with skin-lightening cosmetic products [J]. Clin Toxicol 49(10):886–891
47. Bellinger DC, Stiles KM, Needleman HL (1992) Low-level lead-exposure, intelligence and academic-achievement – a long-term follow-up-study [J]. Pediatrics 90(6):855–861
48. Vicente E, Boer M, Netto C et al (2004) Hippocampal antioxidant system in neonates from methylmercury-intoxicated rats [J]. Neurotoxicol Teratol 26(6):817–823
49. Peng Y, Hu J, Li Y et al (2018) Exposure to chromium during pregnancy and longitudinally assessed fetal growth: findings from a prospective cohort [J]. Environ Int 121.(Pt 1:375–382
50. Hawkesworth S, Wagatsuma Y, Kippler M et al (2013) Early exposure to toxic metals has a limited effect on blood pressure or kidney function in later childhood, rural Bangladesh [J]. Int J Epidemiol 42(1):176–185
51. Smith AH, Marshall G, Yuan Y et al (2006) Increased mortality from lung cancer and bronchiectasis in young adults after exposure to arsenic in utero and in early childhood [J]. Environ Health Perspect 114(8):1293–1296
52. Tsuji JS, Garry MR, Perez V et al (2015) Low-level arsenic exposure and developmental neurotoxicity in children: a systematic review and risk assessment [J]. Toxicology 337:91–107
53. Wasserman GA, Liu XH, Parvez F et al (2011) Arsenic and manganese exposure and children's intellectual function [J]. Neurotoxicology 32(4):450–457
54. Wright RO, Amarasiriwardena C, Woolf AD et al (2006) Neuropsychological correlates of hair arsenic, manganese, and cadmium levels in school-age children residing near a hazardous waste site [J]. Neurotoxicology 27(2):210–216
55. Rosado JL, Ronquillo D, Kordas K et al (2007) Arsenic exposure and cognitive performance in mexican schoolchildren [J]. Environ Health Perspect 115(9):1371–1375
56. Yu ZM, Dummer TJ, Adams A et al (2014) Relationship between drinking water and toenail arsenic concentrations among a cohort of nova scotians [J]. J Expo Sci Environ Epidemiol 24(2):135–144
57. Calderon J, Navarro ME, Jimenez-Capdeville ME et al (2001) Exposure to arsenic and lead and neuropsychological development in mexican children [J]. Environ Res 85(2):69–76
58. Wasserman GA, Liu X, Loiacono NJ et al (2014) A cross-sectional study of well water arsenic and child iq in Maine schoolchildren [J]. Environ Health Glob Access Sci Sour 13(1):23
59. Wasserman GA, Liu X, Parvez F et al (2007) Water arsenic exposure and intellectual function in 6-year-old children in araihazar, Bangladesh [J]. Environ Health Perspect 115(2):285–289

Pesticides: Environmental Exposure, Toxicity, and Mechanism

Ranran Luo and Chunhua Wu

Abstract Pesticides are substances that are meant to control pests or weeds. The production and use of synthetic pesticides have been increasingly growing in recent years worldwide. Pesticides are designed to be toxic to specific groups of organisms, which bring benefits to humans. However, they have considerable adverse effects on other living creatures as well as environmental media including air, water, and soil. Pesticides can enter the human body via the skin, the mouth, and the lungs, thereby leading to diverse toxicity actions on humans. This chapter provides a brief overview of classification, toxicity, health effects, and environmental exposure of pesticides, as well as toxicity and mechanism of three major pesticides (organophosphorus, pyrethroid, and carbamate pesticides).

Keywords Pesticides · Environmental exposure · Toxicity · Toxicity mechanism

1 Introduction

Pesticide is a substance that can kill a pest. Since the number of substances with various chemical and biological subgroups that fall into the major descriptive class of pesticides is enormous, it is hard to define the wide variety of pesticides without overlap. Generally, the broad definition of pesticides refers to substances that control organisms (such as insects, fungi, plants, microorganisms, nematodes, etc.) which can destroy plant life and interfere with the food chain and which act as vectors for disease organisms to man and animals [1]. The Food and Agriculture Organization (FAO) has defined pesticide as: any substance or mixture of substances in order to prevent, destroy, or control any pest, including vectors of human or animal diseases, unwanted species of plants or animals, causing harm during or inter-

R. Luo · C. Wu (✉)
School of Public Health, Fudan University, Shanghai, China

Key Laboratory of Public Health Safety, Ministry of Education, Fudan University, Shanghai, China
e-mail: chwu@fudan.edu.cn

Y. Zhang (ed.), *Emerging Chemicals and Human Health*,
https://doi.org/10.1007/978-981-32-9535-3_7

fering with the production, processing, storage, transport, or marketing of food, agricultural commodities, wood and wood products or animal feedstuffs. The term includes substances that may be administered to animals for the control of insects, arachnids, or other pests in or on their bodies, and substances that are intended for use as a plant growth regulator, defoliant, desiccant, or agent for thinning fruit or preventing the premature fall of fruit, also used as substances applied to crops either before or after harvest to protect the commodity from deterioration during storage and transport.

The pesticides can be classified in many ways in terms of use, toxicity, mode of entry, target organisms, chemical structure, and formulations. According to chemical structure, pesticides can be classified into organochlorides, organophosphates, pyrethroids, carbamates, formamidines, organofluorine, and so on. Specifically, organophosphates are the esters of derivatized phosphoric acid, thiophosphoric acid, and dithiophosphoric acids which are called phosphates, thiophosphates, and dithiophosphates, respectively. Pesticides can also be categorized into insecticides, fungicides, herbicides, rodenticides, and growth regulators on the basis of the target organisms.

The production and use of synthetic pesticides have been increasingly growing in recent years worldwide. Although the increased use of pesticides may benefit agriculture, they are potential sources of environmental pollution, and have adverse effects on ecological environment and human health. Presently, the exposure to pesticides is very extensive and ubiquitous. Occupational exposure frequently occurs in the process of production, transportation, storage, and use of pesticides, while general population can be exposed to pesticides by contacting contaminated products, water, soil, etc. Pesticides can enter the human body via the skin (contact), the mouth (ingestion), and the lungs (inhalation) [2]. Many studies have found different categories of pesticide toxicities like carcinogenicity, mutagenicity, neurotoxicity, immunotoxicity, reproductive toxicity, developmental toxicity, and metabolic toxicity [3, 4]. Besides, toxicity of pesticides may be divided into three main types, i.e., acute, subchronic, and chronic, based on the exposure time to the pesticide and how rapidly the toxic symptoms develop [2].

Pesticide exposure has been linked with various diseases including cancer, hormone disruption, asthma, allergies, and hypersensitivity [5]. Growing evidence also exists for the adverse effects of pesticide exposures relevant to birth defects, reduced birth weight, fetal death, etc. [3] Children during early life are particularly more susceptible to pesticides than adults due to their immaturity of physiological functions and potentially enhanced target organ sensitivity. Therefore, children may be affected even at very low environmental exposure levels.

As pesticides are one of the few toxic substances released deliberately into the environment to kill living organisms, they are universally present in the environment. Moreover, some organochlorides pesticides (e.g., aldrin, chlordane, chlorophenothane (DDT), dieldrin, endrin, heptachlor, and hexachlorobenzene) have served as persistent organic pollutants (POPs), which resist long-term degradation and thus remain in the environment for many years. Moreover, as such compounds have the ability to bioaccumulate and biomagnify, they can be bioconcentrated by up to thousandfold compared to the initial concentration. Repeated application of pesticides leads to loss of biodiversity and increased pest resistance, while its effects on other species facilitate pest resurgence. It is estimated that more than 95% of applied pesticides have the potential to impact nontarget organisms and to become widely dispersed in the environment [6].

Environmental pollution of pesticides is mainly present in three environmental media—air, water, and soil. Airborne pollution of pesticides may occur through pesticide drift (e.g., aerial spray drift) and postapplication volatilization. Even indoors, air currents created by heating, cooling, and ventilation systems can be a mechanism for the spread of pesticides used in indoor environments. Pesticide residues have been found in rain and groundwater as well. Degradation and sorption are both factors that influence the persistence of pesticides in soil. The impacts of some pesticide compounds may last for decades, adversely affecting soil conservation and reducing biodiversity in soil and soil quality [5].

Considering the variety of pesticides, this chapter mainly introduces three major pesticides (organophosphorus, pyrethroid, and carbamate pesticides) and their toxicity mechanisms.

2 Organophosphorus Pesticides (OPs)

The term "organophosphate" is often used as a global synonym for the entire class of OP compounds, including insecticides and nerve agents. General structural formula of OPs is displayed in Fig. 1.

R_1 and R_2 are aryl or alkyl groups, which are bonded to the phosphorus atom directly or through an oxygen or sulfur atom. In some cases, R_1 directly bonds to a phosphorus atom, and R_2 bonds to an oxygen or sulfur atom. In phosphoramide, at least one NH_2 group is present, and double bond is formed through an oxygen or sulfur atom bonding with phosphorus. The X group may be halogen, aliphatic, aromatic, or heterocyclic groups which are also bonded to the phosphorus atom via an oxygen or sulfur atom [7]. When OP is hydrolyzed by phosphotriesterase or interacts with a protein target, the leaving group is released from the phosphorus atom.

OPs are primarily neurotoxicants and act by poisoning the nervous systems of the target species. There are three phases of neurological manifestation induced by OPs: acute cholinergic crisis (ACC), intermediate syndrome (IMS), and organophosphate-induced delayed polyneuropathy (OPIDP) [5]. ACC develops in

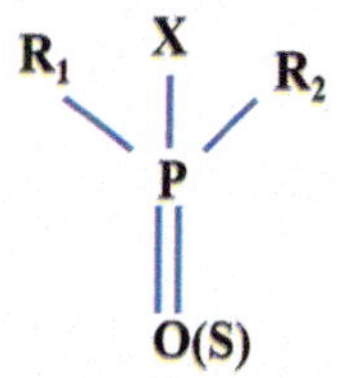

Fig. 1 General structural formula of OPs

the first minutes to hours of OPs poisoning. Due to the anticholinergic action, it affects muscarinic and nicotinic peripheral receptors and the central nervous system. IMS appears after ACC and before OPIDP. IMS is characterized by paralysis of the respiratory and limb girdle muscles, as well as muscles supplied by motor cranial nerves. OPIDP is characterized by degeneration of long axons in the central and peripheral nervous system and ataxia and paralysis that emerges within 2–4 weeks after exposure to certain OPs.

The main mechanism of OPs is to inhibit the function of a group of enzymes named cholinesterase (ChE), causing excessive acetylcholine accumulation at cholinergic effector cells, skeletal neuromuscular junctions, autonomic ganglia, and in the central nervous system (CNS), which finally results in disturbances of nerve conduction function [8, 9]. Certainly, OPs can also produce corresponding toxic effects through other mechanisms. For example, OPs can directly act on cholinergic receptors, other esterases, and even directly affect myocardial cells to cause myocardial damage [10].

2.1 *Neurotoxic Effects of Organophosphorus Pesticides*

2.1.1 Acetylcholinesterase inhibition

Cholinesterases belong to a family of carboxylesterase enzymes that catalyze the hydrolysis of the acetylcholine back to choline and acetic acid, an essential process for cholinergic neurons to restore [9]. Acetylcholine is a neurotransmitter released by the cholinergic and central nervous systems. Acetylcholine is present in all autonomic ganglia, most autonomic innervation organs, neuromuscular junction, and most synapses in the central nervous system [7]. Acetylcholine is used to synthesize choline from choline and acetic acid under the action of choline acetyltransferase in cholinergic neurons. During the neurotransmission, acetylcholine is released into the synaptic cleft and bonded to its receptors on postsynaptic neurons, causing them to depolarize. It must be quickly removed from the synaptic cleft to prevent lateral diffusion and frequent activation of adjacent receptors. The efficient removal is due to the presence of cholinesterase in presynaptic and postsynaptic cell membranes. After acetylcholine is decomposed, the acetate diffuses away and choline is brought back to the presynaptic nerve terminal to synthesize more acetylcholine.

There are two types of cholinesterases, namely, acetylcholinesterase (AChE), also known as erythrocyte cholinesterase or true cholinesterase, and butyrylcholinesterase (BChE) or pseudocholinesterase. AChE is found mainly in the nerve system and red blood cell membranes and plays an important role in terminating the action of acetylcholine at synapse. BChE is primarily synthesized in the liver and is also present in the plasma. It can hydrolyze butyrylcholine quickly, but its hydrolysis of acetylcholine and propionylcholine is weak and still unclear.

AChE is responsible for terminating neurotransmission at cholinergic nerve endings by rapid hydrolysis of the neurotransmitter acetylcholine. AChE has two active sites including the anionic subsite and the esteratic subsite [7]. The anionic subsite binds the positively charged nitrogen of acetylcholine to form choline, and the esteratic subsite combines with the acetyl carbon atom of acetylcholine to form acetylcholinesterase. Acetylcholinesterase is hydrolyzed by acetylhydrolase to change an acetyl group into acetic acid. Acetic acid diffuses away and cholinesterase restores its original state to repeat its function.

When cholinergic nerves are stimulated, nerves terminal will release acetylcholine. Then acetylcholine acts on the effector to produce a series of neurotoxic symptoms. The symptoms can be divided into three types: muscarinic symptoms, nicotinic symptoms, and central nervous system symptoms.

OPs attach to the esteratic site of AChE through phosphorylating the serine hydroxyl group, forming phosphoryl cholinesterase [11]. In this way, the hydrolytic activity of the enzyme is inhibited because the active site for physiological substrate has been occupied. What is more, the stability of the bond between phosphorus atom and esteratic site of the enzyme is much more than the bond between the carbonyl carbon of acetylcholine and the same site. Phosphoryl cholinesterase cannot hydrolyze acetylcholine, so that acetylcholine released from the cholinergic nerve terminal accumulates in corresponding organs or tissues, causing symptoms similar to hyperexcitability of cholinergic nerves, such as muscarinic symptoms, nicotinic symptoms, and central nervous system symptoms. However, excessive acetylcholine can cause muscle contraction, glandular secretion, and depolarization of the neuronal membrane endplates, resulting in muscle weakness and paralysis.

2.1.2 Organophosphate-Induced Delayed Polyneuropathy (OPIDP)

It is found that neuropathy target esterase (NTE) is involved with the development of OPIDP. NTE is an integral membrane protein in vertebrate neurons, whose activity depends on lipid content. It is present in the endoplasmic reticulum of neurons but not in glial cells. NTE has important roles in mammalian development since it is essential for mammalian embryo development. NTE regulates phospholipid metabolism and is known to be a phospholipase B. NTE can hydrolyze a variety of carboxylic acid esters. When reacting with OPs, NTE is gradually suppressed with time and temperature. In this reaction, phosphorylated NTE may lose an alkyl group bound to phosphorus ("aging" reaction) [12].

Some evidence show that inactivation of NTE may reduce the degradation of phosphatidylcholine to glycerophosphocholine [12]. The lack of NTE activity may lead to abnormal accumulation of phosphatidylcholine-containing membranes in cells, thereby interfering with normal membrane lipid homeostasis and fluidity, and finally affecting the formation of neurites. As NTE has been proved to hydrolyze lysolecithin, the delayed initiation of neurites in NTE-deficient cells may result from the abnormal accumulation of lysophospholipids. In addition, NTE is involved in cell membrane transport and cell signaling pathways between neurons and glial cells and facilitates the transport of macromolecules from the neuron to the distal ends of long axons. NTE may be also associated with the regulation of calcium entrance into cells that are responsible for the maintenance of normal function of calcium channels, and increasing calcium-activated neutral protease activity may trigger OPIDP.

It is generally considered that the main mechanism of OPIDP development is the inhibition of NTE by phosphorylation. A sufficient amount of NTE must be inhibited for OPIDP to develop [13]. Although the inhibition of NTE is necessary for the development of OPIDP, its separate presence is not sufficient for the development of neuropathy. The development of OPIDP mostly requires "aging" inhibition of the NTE enzyme, with the reasons unclear. However, some studies indicate that aging of NTE is not a necessary condition for the development of OPIDP. Some OPs that inhibit NTE without aging have been shown to cause neuropathy at very high doses, almost completely inhibiting NTE. Metabolic abnormalities and nerve conduction dysfunction, induced by inhibited NTE, may also be involved in pathogenesis of OPIDP.

2.1.3 Intermediate Myasthenia Syndrome

The pathogenesis of intermediate myasthenia syndrome (IMS) has not yet been clear [14]. Its nervous system symptoms seem to reflect a pathophysiological dysfunction of the neuromuscular junction due to the accumulation of acetylcholine in the synaptic cleft. Other mechanisms include striated muscle necrosis, persistent AChE inhibition, decrease of serum potassium levels, and oxidative stress damage.

2.1.4 Neurodegenerative Diseases

Aging, which can be classified as neurodegenerative diseases, is often related to a wide variety of clinical and pathological conditions. Typical examples of these diseases include Parkinson's disease (PD), Alzheimer's disease (AD), or amyotrophic lateral sclerosis (ALS). Although the etiology of these diseases is unknown, both environmental and genetic factors may play an essential role. Oxidative stress and neuroinflammatory processes are prominent in most neurodegenerative diseases [11].

Evidence in experimental and epidemiological studies indicates that neurodegenerative diseases are associated with OPs exposure [9]. However, the exact etiology of these health problems is still unclear; more mechanistic evidence is required to explore their relationships with OPs.

2.2 *Toxic Effects on Biomembrane*

Since OPs are generally fat-soluble, they may affect the structure of biomembrane lipids, changing membrane structure or its physiological functions. These toxic effects may be related to noncholinergic mechanism of OPs. Some studies have demonstrated that the fat-soluble parathion has toxic effects on cell membranes, while monocrotophos with strong water solubility is not clearly toxic to cell membranes.

Parathion can inhibit the activity of calcium pumps on human erythrocyte membrane, as well as rat erythrocyte membrane and brain synaptic knob membrane [15]. In addition, the binding site of parathion and calcium pump is different from the site of ATP combined with calcium pumps. The inhibition of parathion on calcium pump is independent of the presence or absence of calmodulin. However, monocrotophos, with both fat solubility and water solubility, does not inhibit calcium pumps, indicating that the chemical structure and physical properties of organophosphorus pesticides can influence their affinity with membranes. It is also found that parathion can directly act with the membrane, interfere with the active transport of calcium ions, prevent the outflow of calcium ions from red blood cells, and damage the intracellular calcium homeostasis. Parathion can also induce a decrease of lipid fluidity in human ghost cell membrane and rat brain synaptic knob membrane, while monocrotophos does not have these effects.

2.3 *Effect of OPs on Phase II Metabolic Enzymes*

Glutathione S-transferases (GSTs) comprise a family of eukaryotic and prokaryotic phase II metabolic isozymes, involved in the metabolic transformation of the organic phosphorus compound. Chlorpyrifos (CPF) is a broad-spectrum organophosphate insecticide, the metabolic transformation process of which affects the toxicity to the body. CPF has an isozyme selective effect on GST in the liver and its effect is associated with the dose of CPF. Some studies also show that the effect of CPF on GST in rat brain tissue has both brain selectivity and isozyme selectivity [15].

2.4 *Reproductive and Developmental Toxicity*

The US EPA defines EDCs as "exogenous agents that interfere with the synthesis, secretion, transport, binding, action or elimination of natural hormones responsible of metabolism, reproduction, development and behavior." Many pesticides have been classified as EDCs; among them are various OPs and carbamates pesticides, fungicides, and herbicides. The toxicity underlying their action may be correlated with the inhibitory effects of OPs on brain AChE, which may affect the regulation of hypothalamic function by altering the frequency of GnRH pulses that play an important role in determining the output of LH and FSH from the pituitary [16].

Several epidemiological studies have shown associations between pesticides exposure and change of semen quality, which directly or indirectly influence fertility and reproduction in agricultural workers [17]. Many occupational studies indicate that exposure to OPs is associated with sperm DNA damage and oxidative stress.

The neurotoxicity of OPs has a serious damage to the development of nervous system of fetuses and developing infants, because of their immature metabolism and elimination systems [18]. Evidence from in vitro and animal models and epidemiological studies has shown associations between OPs exposure and neurodevelopmental toxicity [19]. The mechanism of effects on neurologic development is complex, including well-known cholinesterase inhibition, excitatory amino acid transmitter system, dopaminergic neurotransmitter system, serotonin transmitter system, extracellular regulated protein kinase/cAMP response element-binding protein (ERK/CREB) signaling pathway, and oxidative stress [19, 20].

2.5 *Carcinogenicity, Teratogenicity, and Mutagenicity*

Some epidemiological studies and animal experiments suggest that OPs are linked with non-Hodgkin's lymphoma, leukemia, and lung cancer [21]. The mutagenicity of OPs has been reported both at home and abroad. For instance, dimethoate can increase the occurrence of sister chromatid exchange (SCE) in mouse bone marrow cells in a dose-response relationship. Besides, an epidemiological study showed that the chromosomal aberration rate and SCE in peripheral blood cells of dichlorvos production workshop workers were significantly higher than those of the control group [15]. The specific mechanisms of these toxicities are unclear. Oxidative stress may be one of the mechanisms. Reactive oxygen species (ROS), induced by OPs, attack lipids, proteins, and DNA, causing oxidation and membrane damage, enzyme inactivation, DNA damage, and cell death [9]. Additionally, the genotoxicity of OPs may be related to their electrophilicity [21]. Several OPs are electrophiles capable of interacting with the nucleophiles in organisms by alkylation. The alkylating compounds are frequently mutagenic or carcinogenic to the exposed organisms, through the induction of severe cellular stress including chromosome aberrations, SCE, gene mutations, and cell killing.

3 Pyrethroid Pesticides

Pyrethroids are a class of synthetic insecticides. Their chemical structure is similar to naturally occurring pyrethrins, found in the flowers of *Chrysanthemum cinerariaefolium*. The basic pyrethroid structure is composed of an acid and an alcohol moiety, with an ester bond. Pyrethroids are among the most frequently used pesticides and are found in many formulations used in agriculture and indoor environments.

Historically, pyrethroids are classified into two subclasses (Types I and II), according to chemical structure and the production of either the T (tremor) or CS (choreoathetosis with salivation) poisoning syndrome following intravenous or intracerebral administration to rodents [22]. Type II pyrethroids contain the α-cyano-3-phenoxybenzyl moiety (e.g., cypermethrin, deltamethrin, and fenvalerate), whereas Type I compounds comprise a wide structural variety of compounds lacking the α-cyano-3-phenoxybenzyl group (e.g., pyrethrin I, resmethrin, and permethrin) (Fig. 2) [23]. The Type I/II nomenclature has been widely adopted in

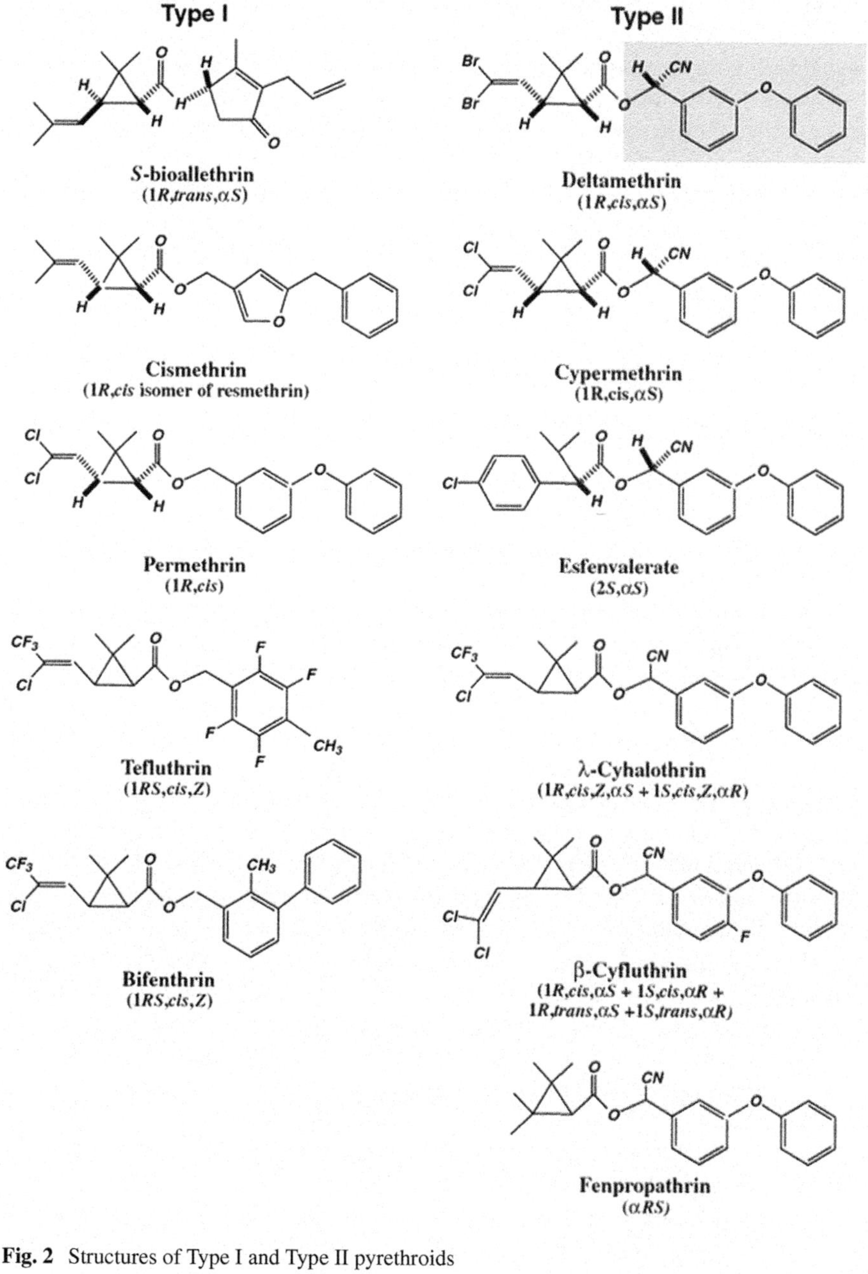

Fig. 2 Structures of Type I and Type II pyrethroids

the literature and is often used in a manner parallel to the T/CS nomenclature, so that Type I compounds are generally considered to produce the T syndrome of intoxication and Type II compounds are considered to produce the CS syndrome. However, the correspondence between the Type I/II (structural) and T/CS (symptomatic) taxonomies is not perfect, because some Type II structure compounds may produce either the T syndrome of intoxication or the signs of intoxication that exhibit elements of both the T and CS syndromes. Generally, references to Type I and Type II compounds only denote the structural classes.

Occupational exposure to pyrethroid pesticides is mainly found in the cultivation and harvesting of pyrethrum, the production, disassembly, transportation, storage, sale, and use of pyrethroid pesticides. During field application of pesticides, violation of operational regulations is the main cause of occupational acute poisoning. Exposure of general population to pyrethroids mainly occurs through ingestion of residues in food, or ingestion of or dermal contact with contaminated house dust or surface-adhering particles, following domestic use [24]. Pyrethroid pesticides are low toxic and safe to use indoors.

Systemic poisoning of pyrethroid pesticides often occurs for mistaken ingestion and lack of appropriate protective conditions. It is difficult to control skin contamination when spraying pyrethroid pesticides. Systemic poisoning is usually characterized by central nervous system symptoms and is easily confused with other pesticides such as anticholinesterase pesticide poisoning, since pyrethroids can also cause salivation and excessive excitement. In addition, solvents used in pyrethroid pesticides, such as xylene, may induce nervous system symptoms, which make pyrethroid pesticides poisoning more complicated.

Although pyrethroid pesticides are mostly low toxic to humans, protective measures should not be neglected. Pyrethroid pesticides are neurotoxic agents that interfere with nerve conduction and induce poisoning, even affects neurological development of children [25, 26]. The nerve poisoning symptoms include three types, T syndrome, CS syndrome, and mixed and intermediate poisoning. T syndrome mainly induced by Types I pyrethroids is the simplest poisoning syndrome. Its poisoning symptoms are similar to the symptoms caused by DDT. Type I pyrethroid poisoning involves the progressive development of the following symptoms: systemic small tremors, extraordinary startle response, uncoordinated twitching of back muscles, the excessive excitement, and eventually death. CS syndrome induced by Type II pyrethroids is more complex, which involves broader tissues. Type II pyrethroid poisoning of human is characterized by sensory abnormalities, dizziness, nausea, burnout, and fasciculation. More serious poisoning manifestations include abdominal pain, nausea and vomiting, phlegm and pulmonary edema, angular arch reversal, disease onset, and coma. Some pyrethroids can produce complex and mixed manifestations of Type I and Type II combination poisoning, reflecting the overlap and transitional state of T and CS syndromes [15].

Additionally, paresthesia is frequently reported symptom in occupational exposure studies, characterized by numbness, itching, burning, or tingling of the skin following dermal exposure to a pyrethroid [22]. Paresthesia seems to be a local effect, and only occurs at the site of dermal exposure. It is not associated with the appearance of a rash or other signs of classical skin irritation and any signs of systemic intoxication.

3.1 Voltage-Sensitive Sodium Channels

Voltage-sensitive sodium channels are the main site on that pyrethroids act, and many studies reveal that it is also an important site of pyrethroid action in mammals. The effects of pyrethroids on sodium channel function are in accord with the mechanism of enhanced neuronal excitation, and these effects also reflect the stereospecificity of the effects associated with the expected toxicology. Moreover, the differences observed in the modification of sodium channels by α- and noncyano pyrethroids have consistency with the differences in the disruption of nerve action potentials resulted from members of these two structural groups and with the production of CS and T syndromes. Most Type I compounds induce explosive discharges in invertebrate neurons after a single stimulus and rapidly attenuate sodium tail currents under voltage-clamping conditions, resulting in a poisoned T syndrome in rodents. Most Type II compounds, under voltage-clamping conditions, result in the use-dependent nerve block due to a decrease in resting membrane potential of invertebrate neurons and a long-lived sodium tail current, finally causing the CS syndrome. A few compounds have an intermediate effect on neuronal excitability and tail current decay, and produce signs of poisoning, including elements of T and CS syndrome. However, biophysical experiments have shown that two distinct functional classes, α- and noncyano pyrethroids, represent a simplification. Because biophysical studies identify the spectrum of effects of pyrethroids on sodium channel function, and these effects are associated with the development of symptoms of poisoning that are intermediate between T and CS syndromes.

3.2 Voltage-Sensitive Calcium Channels

Evidence from biochemical and electrophysiological studies in vitro suggests that pyrethroids have influence on calcium channel function. Pyrethroids are antagonists of T-type calcium channels in vertebrate tissues and agonists of T-type channels in insect muscle and in the ciliary membrane of *Paramecium tetraurelia*. The effect of pyrethroids on N-type calcium channels is also indirectly associated with incomplete elimination of pyrethroid-stimulated neurotransmitter release under some experimental conditions, which eliminates the effect on sodium channels. However, the direct action of pyrethroids on N-type calcium channel currents has not been found. The effect of pyrethroidsin these assays shows appropriate stereospecificity for the pair of detected isomers. Biochemical and electrophysiological evidence indicates that L-type calcium channels are relatively insensitive to pyrethroids, but the sensitivity of other types of calcium channels to pyrethroids has not been studied. Although a large number of calcium channel α1 subunits have been cloned, only some of these express their functions. The determination of pyrethroids on detected calcium channel isoforms, similar to determination of cloned sodium channel isoforms, has not been performed. Although the effect of pyrethroids on N-type calcium channels is in line with the release of large amounts of neurotransmitters caused by pyrethroid in vivo, the toxicological significance of pyrethroids on calcium channels is not clear. The connection between effects on calcium channel observed in vitro and pyrethroid poisoning is unclear, either at the intact neurological levels or at the whole animal. Furthermore, the effects of pyrethroids seem not to be associated with the action of sodium channels, neither in vitro nor in vivo. Thus, although the effects on calcium channels may result from at least some pyrethroids to some extent, there is currently no basis to confirm that the effect on calcium channels is an essential mechanism of pyrethroid poisoning.

3.3 Voltage-Sensitive Chloride Channels

The effect of pyrethroids on voltage-specific chloride channel is assumedly limited to α-cyano structure group, but this conclusion is on the basis of experiments with deltamethrin as the sole example of the α-cyano group and cismethrin as the sole example of the noncyano group. The blockade of voltage-dependent chloride channels caused by deltamethrin results from the excitation of the nerve and muscle in combination with other neuroexcitatory effects; thus, it has consistency with the overall neuroexcitatory effects of pyrethroids. So far, the stereospecificity of pyrethroids acting on this target is unclear.

Additional evidence for the toxicological relevance of pyrethroid actions on voltage-sensitive chloride channels comes from pharmacological studies in vivo. For example, ivermectin and pentobarbital can selectively activate voltage-dependent chloride channels, antagonize the symptoms of deltamethrin poisoning, and reduce mortality with deltamethrin used but have no similar effect on cismethrin intoxication. The effects of these drugs on salivation may due to the antagonistic effect on the chloride channels in peripheral nervous system, while the symptoms of exercise-induced poisoning caused by deltamethrin may due to the antagonistic effect on the chloride channels in central nervous system. Although the limited number of pyrethroids is used in these studies, the results indicate that the activity of voltage-sensitive chloride channels may lead to some symptoms of poisoning related to CS syndrome.

Besides, some evidence show that pyrethroids selectively regulate cardiac chloride conductances, suggesting these insecticides play a completely new role on chloride channels. These effects include activation and inhibition of different chloride conductances, at least in one example by the same pyrethroid. The selective action of pyrethroids on chloride transport mechanisms indicates that they are assumedly valuable probes for the pharmacological dissection of these processes, with no evidence at present confirming these effects on cardiac chloride conductances play a role in pyrethroid poisoning.

3.4 GABA Receptors

A large number of biochemical evidence documents the ability of α-cyano pyrethroids to bind to and block GABA receptors in mammalian brain preparations [22]. As an indirect neuroexcitatory effect including the removal of inhibitory neuronal input, blockade of GABA receptors is the established mode of action for convulsive drugs. From a functional perspective, the action on GABA receptors is in accordance with the neuroexcitatory signs of pyrethroids poisoning in vivo.

The action of pyrethroids on GABA receptors is somewhat stereoselectively toxic to isomers of α-cyano compounds, while these do not show the absolute stereospecificity predicted by structure-toxicity relationships. In experimental studies, the effects of GABA receptors and sodium channels can be detected in the same formulation, while GABA receptor blockade is not observed when pyrethroids disrupt sodium channel function. Information about in vitro biochemical effects of pyrethroids on GABA receptors with intoxication in vivo is very limited. The CS syndrome following intracerebral administration of pyrethroids to mice was similar to the signs of intoxication produced by the intracerebral administration of picrotoxinin. Moreover, the relatively low potency and incomplete stereospecificity of pyrethroids in functional assays do not support that the GABA receptor is an important target site for the production of CS syndrome.

3.5 Peripheral-Type Benzodiazepine Receptors

Many pyrethroids can inhibit the binding of the ligand Ro5–4864 to peripheral-type benzodiazepine receptors [22]. Considering that Ro5–4864 is a convulsant, other convulsants may interfere with the bond between Ro5–4864 and its bonding site. Moreover, the inhibition of Ro5–4864 binding shows stereospecificity to some extent, which is in accord with the stereospecificity of pyrethroid poisoning.

Evidence for the effects of peripheral benzodiazepine receptors during poisoning comes from pharmacological studies in vivo. When the dose of pyrethroids is lower than the poisoning dose, it can potentiate the convulsant actions of pentylenetetrazole by that way similar to Ro5–5864, and these proconvulsant effects are blocked by PK11195, a specific antagonist of Ro5–4864. These findings indicate pyrethroids can interact with peripheral benzodiazepine receptors, but this interaction cannot be the primary mechanism of poisoning for it only occurs at the dose that are without effect outside of the proconvulsant experimental model. What is more, the proconvulsant effects of cismethrin and deltamethrin have not been replicated by other researchers.

Although results are conflicting in vivo, the effect of pyrethroids on the peripheral benzodiazepine receptor may lead to pyrethroid poisoning by enhancing convulsions arising caused by nerve excitation effects of other targets. The ability to stabilize the incubation period of pyrethroid poisoning can support this interpretation. Originally, this effect was explained as evidence of the effect of pyrethroids on GABA receptors; but diazepam has a high affinity with GABA receptors and peripheral benzodiazepine receptors in rodents. The effect of diazepam on the latency of pyrethroid poisoning is more likely due to the antagonism at the peripheral-type benzodiazepine receptor than the GABA receptor. So some pyrethroids may be used as proconvulsants at peripheral-type benzodiazepine receptors at toxicologically relevant doses. However, the relationship between this effect and the main neuroexcitatory effects of pyrethroids and its contribution to the signs of pyrethroid poisoning remains unclear.

3.6 *Ion Channel and Receptor Regulation*

Several evidence exhibit the ability of pyrethroids to modify second messenger pathways involved in ion channel and neurotransmitter receptor regulation [22]. Additionally, pyrethroids are likely to have specific effects on the pattern of protein phosphorylation in neurons. Although some of these effects reflect neuronal responses to the neuroexcitatory effects of pyrethroids rather than primary mechanisms of toxicity, others may reflect a more specific effect of pyrethroids. Given that $\beta\gamma$ subunits of G proteins can directly modulate of N-, P-, and T-type calcium channels, the bond between pyrethroids and the $\beta\gamma$ subunits implied by the results of photoaffinity labeling experiments could contribute to effects on a number of macromolecules, including sodium and calcium channels. Currently, these effects are only found in observations from in vitro experiments and are not directly associated with pyrethroid poisoning in vivo.

3.7 *Napoptosis*

Currently, there are two recognized mechanisms of cell death: necrosis and apoptosis. Necrosis is a mechanism in which the damage caused by internal and external agents reaches a certain intensity, and the cell death occurs after a certain period of time. Apoptosis is active spontaneous death related to genetic mechanism, and it has a series of characteristic manifestations of morphology and biochemistry. Apoptosis is different from cell necrosis, which is associated with cell signal transduction. Specifically, apoptotic signals act on cell membrane receptors, changing intracellular messengers and consequent gene expression, and ultimately leading to cell death.

There are many methods for studying apoptosis [15]. The first is flow cytometry. The principle is that apoptotic cells, with nuclear DNA splitting into many small fragments, may be lost through the nuclear membrane, leaving only the large fragment of DNA. The missing DNAs of the cells form a peak (sub-G1 peak), whose DNA content is less than 2 N. The sub-G1 peak is used to judge whether apoptosis occurs or not. Studies have shown that after being treated with deltamethrin and permethrin, significant sub-G1 peaks appear in the cortex, hippocampus, and cerebellum of rat brain tissues, indicating that apoptotic cells are significantly increased, and the average fluorescence intensity of DNA in the nucleus is lower than the control group. The results show that pyrethroid pesticides can induce apoptosis of nerve cells. The second is end labeling method. The principle is that when apoptosis occurs, the DNA strand is cut by activated endogenous endonuclease into a number of fragments about 180 bp, which contain 3′-OH ends. Adding the labeled nucleotide mixture, under the action of terminal converting enzyme or DNA polymerase, these fragments can be judged for apoptosis by enzymatic color development or fluorescent antibody. Third is DNA gel electrophoresis. Another feature of apoptosis is that the DNA strand is cut by activated endogenous endonuclease into a number of fragments, the length of which is 180–220 bp or multiples. These fragments show a typical "ladder" strips after agarose gel electrophoresis. That is, the results of DNA electrophoresis of apoptotic cells are fragmented. Fourth is HE staining and Nissl staining. The two staining methods are commonly used in neuronal apoptosis studies. In rat cerebral cortex and hippocampus following deltamethrin exposure, for instance, after Nissl staining, many Nissl bodies are shrinking and dark purple. Various experimental methods and techniques have confirmed that deltamethrin can induce neuronal apoptosis. Using techniques such as in situ end labeling and HE staining, apoptotic cells can be detected in in vitro experiments.

4 Carbamate Pesticides

Carbamates are organic compounds derived from carbamic acid (NH_2COOH) or thiocarbamic acids. The structure of biologically active carbamates is displayed in Fig. 3. As depicted in Fig. 3, X can be oxygen or sulfur, R_1 and R_2 are usually organic or alkyl substituents, but R_1 or R_2 may also be hydrogen, and R_3 is mostly an organic substituent or sometimes a metal [7]. Carbamate compounds are used as

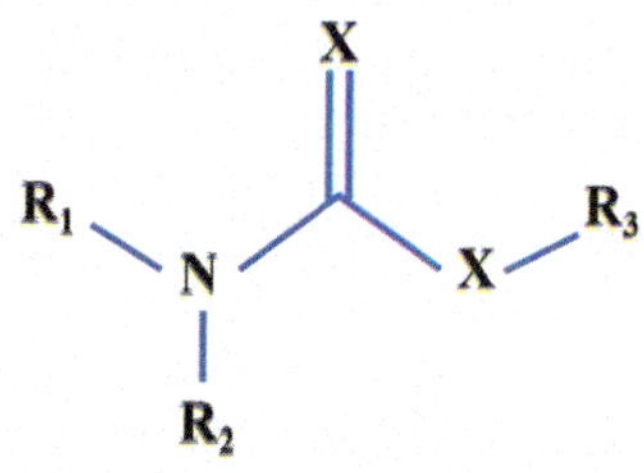

Fig. 3 General chemical structure of biologically active carbamates

fungicides, insecticides, and herbicides in agriculture and are identified as the second group of pesticides inhibiting cholinesterases.

Their acute toxicity to humans is generally low, but they may irritate the skin, eyes, and throat, leading to sneezing and coughing. Carbamates can reversibly inhibit AChE, so it has an important application in human medicine. As the secondary metabolite in the plant, physostigmine is widely used in the treatment of myasthenia gravis. In addition, they can be used as therapeutic drugs in other diseases (such as AD, glaucoma, Lewy bodies, and Parkinson's disease). As a potent AChE inhibitor, this therapeutic agent can reduce ACh hydrolysis rate, and thereby increases its level in damaged neurosynaptic clefts improving nerve impulse transmission.

One route of occupational exposure to carbamate pesticides is skin contact. Due to the lipophilic nature of carbamate pesticides, as most formulations contain organic solvents and emulsifiers, they can penetrate the skin quickly and be absorbed into the blood circulation. High temperature and high humidity may increase the absorption. Larger bare skin areas lead to more complete absorption. Additionally, carbamates are easily absorbed by the gastrointestinal system. Nonoccupational exposure is mainly resulted from pesticide residues in food, which is usually below the maximum allowable residue level and theoretically safe to humans. However, if people do not follow the pesticide management rules, pesticide residues can cause food poisoning. Inhalation is another route of exposure. When the pesticide is applied in a high-temperature environment, the exposure of sprayer is greatly increased due to the rapid evaporation of pesticides, even inducing acute poisoning [15].

4.1 Neurotoxicity

Carbamates and OPs are AChE inhibitors. They produce both similar toxic acute effects and symptoms caused by poisoning. The main difference in AChE inhibitory action between OP and carbamate is the stability of the AChE-OP/carbamate complex. The stability of the carbamate-AChE bond is much less than OP-AChE bond, and spontaneous decarboxylation (by slow hydrolysis) occurs after a while, leading to reactivation of AChE. Furthermore, the enzyme inhibitor complex does not undergo aging process like OPs. Nonetheless, the transient inhibition of AChE by carbamates allows ACh to accumulate in the muscarinic and nicotine synapses of the sympathetic and parasympathetic nervous systems, as well as in the neuromuscular junctions of striated and smooth muscles, ultimately leading to a clinical cholinergic crisis.

Furthermore, carbamates, analogously to OPs, reversibly inhibit NTE but, unlike OPs, are not able to dealkylate, i.e., age the inhibited enzyme. So, carbamates are not delayed neuropathy inducers. Moreover, they exhibit protective effects when applied before OPs induced neuropathy. The neuropathic OPs are not able to inhibit and age neuropathy target esterase previously reversibly inhibited by carbamates.

However, carbamates stimulate delayed neuropathy or make it more severe, when they are dosed after applying OP neuropathic doses, inducing promotion of delayed neuropathy.

4.2 Reproductive Toxicity

There is evidence that occupational exposure to endocrine disrupting pesticides, such as OPs and carbamate pesticides, has the potential impact on male reproductive function. Chronic exposure to these pesticides may do harm to sperm chromatin, decrease semen quality, and change male reproductive hormones, causing adverse reproductive health outcomes [16]. However, current experimental evidence for the reproductive toxicity of carbamate insecticides in mammals is not sufficient. The exact mechanism of this action needs more evidence to elucidate.

4.3 Immunotoxicity

It is reported that exposure to carbamate pesticides significantly increases the risk of non-Hodgkin's lymphoma in humans, indicating that carbamate pesticides may impair human immune system. Natural killer cells (NK) and cytotoxic T lymphocytes (CTLS) can provide host defense against tumors. Studies of the effect of carbamate pesticides on NK cells suggest that carbamate pesticides significantly inhibit human NK activity by reducing intracellular perforin, granzyme A, granzyme B, granzyme 3/K, and granulysin levels in NK cells. Additionally, several carbamate pesticides induced apoptosis in human NK cells and T cells, and the apoptosis was mediated by the activation of caspases and the release of mitochondrial cytochrome-c [27, 28].

4.4 Developmental Neurotoxicity

Carbamates poisoning syndrome is usually not as serious as the syndrome caused by OPs poisoning. But carbamates have been confirmed to produce neurotoxic effects, and exposure during development is more harmful than at adulthood, which is similar to other classes of insecticides.

The comparison of rodents different postnatal ages, exposed to several carbamate pesticides, show that younger animals need smaller dose to produce AChE inhibition and cholinergic toxicity. The increased sensitivity can be resulted from a reduced clearance and consequent increased brain carbamates levels in younger animals, which is reported to occur with carbaryl and 1-naphthol. However, whether similar pharmacokinetic differences also occur with other components of the same class and with prolonged periods of exposure is still poorly understood.

The effects of exposure during development are diverse and changes can be found even in adulthood. The deleterious effects of acute toxicity are largely dose-independent; highly toxic compounds, moderately toxic compounds, or low toxic compounds all contribute to neurobehavioral effects with developmental exposure. Some studies show that developmental exposure to different components of the carbamates class can lead to different effects, along with the transient inhibition of AChE, and indicates that the mechanisms associated are not restricted to those involved in the acute toxicity [19, 29].

4.5 *Effect on Target Human Melatonin Receptors*

The nightly release of melatonin from the pineal gland is modulated by biological clocks in the suprachiasmatic nucleus (SCN) of the hypothalamus, transmitting signals to the brain and peripheral target tissues via activation of the G-protein-coupled melatonin receptors MT1 and/or MT_2. Additionally, the activation of melatonin receptors plays a key role in regulating the phase and amplitude of circadian rhythms. Circadian rhythm misalignment is involved in various syndromes such as delayed sleep phase syndrome (DSPS) and seasonal affective disorders (SAD).

Activation of melatonin receptors inhibits the formation of cyclic adenosine monophosphate (cAMP) and regulates phosphorylation of the cAMP-responsive element-binding protein (CREB), protein kinase activity, and ion flux. Therefore, the melatonin receptor system can regulate a series of cellular signaling events that are critical for maintaining rhythmicity and homeostatic balance of various regulatory processes. Abnormal melatonin receptor activation, sensitivity, and trafficking are associated with pathological conditions including diabetes, cardiovascular diseases, and cancer.

Some carbamate insecticides, such as carbaryl and carbofuran, are structural mimics of the neurohormone melatonin and able to bind to melatonin receptors. Exposure to these insecticides can either mimic (MT_2) or alternatively block (MT_1) the effect of endogenous melatonin potentially affecting melatonin receptor-mediated signaling in target tissues. Also, the carbamate insecticides carbaryl and carbofuran are reported to alter the metabolism of serotonin (a metabolic precursor of melatonin) and melatonin [30].

4.6 *Genotoxicity*

The evidence for mutagenicity of carbamate pesticides in animals is not sufficient [31]. Several carbamate pesticides have been identified to have weak mutagenicity, but their exposure doses are all at high levels. Besides, the results are often unrepeatable, or derived from nonstandardized experiments, which are incomparable with conclusions derived from accepted standardized assays.

4.7 Carcinogenicity

Some epidemiological studies have shown that carbamate pesticides are involved in the development of many types of cancer [32]. The occurrence of cancer may be related to the immunotoxicity of carbamate pesticides [33]. However, the evidence for carcinogenicity of carbamate pesticides in animals is not sufficient. The interpretation of the positive results must be cautious because the results may be from unreasonable study design and extraordinarily high exposure dose.

5 Summary

Pesticides bring many benefits to humans in agriculture, industry, and health areas, but their harm to humans, animals, and the environment has always attracted widespread attention. This chapter mainly introduces environmental exposure of pesticides, as well as toxicity and mechanism of three major pesticides (organophosphorus pesticides, pyrethroid pesticides, and carbamate pesticides). In addition, this chapter provides a brief overview of classification, toxicity, human exposure routes, and health hazards of commonly used pesticides.

References

1. Marrs TC, Ballantyne B (2004) Pesticide toxicology and international regulation [M]. Wiley, Chichester
2. Damalas C, Koutroubas S (2016) Farmers' exposure to pesticides: toxicity types and ways of prevention [J]. Toxics 4(1):1
3. Mostafalou S, Abdollahi M (2017) Pesticides: an update of human exposure and toxicity [J]. Arch Toxicol 91(2):549–599
4. Patel S, Sangeeta S (2019) Pesticides as the drivers of neuropsychotic diseases, cancers, and teratogenicity among agro-workers as well as general public [J]. Environ Sci Pollut Res 26(1):91–100
5. Kim K, Kabir E, Jahan SA (2017) Exposure to pesticides and the associated human health effects [J]. Sci Total Environ 575:525–535
6. Simeonov LI, Macaev FZ, Simeonova BG (2013) Environmental security assessment and management of obsolete pesticides in Southeast Europe [M]. Springer, Dordrecht
7. Colovic MB, Krstic DZ, Lazarevic-Pasti TD et al (2013) Acetylcholinesterase inhibitors: pharmacology and toxicology [J]. Curr Neuropharmacol 11(3):315–335
8. Mileson BE, Chambers JE, Chen WL et al (1998) Common mechanism of toxicity: a case study of organophosphorus pesticides [J]. Toxicol Sci 41(1):8–20

9. Mostafalou S, Abdollahi M (2018) The link of organophosphorus pesticides with neurodegenerative and neurodevelopmental diseases based on evidence and mechanisms [J]. Toxicology 409:44–52
10. Joshi P, Manoria P, Joseph D et al (2013) Acute myocardial infarction: can it be a complication of acute organophosphorus compound poisoning? [J]. J Postgrad Med 59(2):142–144
11. Costa LG (2018) Organophosphorus compounds at 80: some old and new issues [J]. Toxicol Sci 162(1):24–35
12. Jokanović M, Kosanović M, Brkić D et al (2011) Organophosphate induced delayed polyneuropathy in man: an overview [J]. Clin Neurol Neurosurg 113(1):7–10
13. Akçay Yalbuzdag Ş (2017) Organophosphate induced delayed neuropathy: a case report [J]. Türk Fiz Tıp Rehab Derg 63(1):88–91
14. Haliga RE, Morarasu BC, Ursaru M et al (2018) New insights into the organophosphate-induced intermediate syndrome [J]. Arch Ind Hyg Toxicol 69(2):191–195
15. Xia S. Pesticide toxicology [M]. Chemical Industry Press, Beijing
16. Miranda-Contreras L, Gomez-Perez R, Rojas G et al (2013) Occupational exposure to organophosphate and carbamate pesticides affects sperm chromatin integrity and reproductive hormone levels among Venezuelan farm workers [J]. J Occup Health 55(3):195–203
17. Recio-Vega R, Carnpo-Gomez G, Borja-Aburto VH et al (2008) Organophosphorus pesticide exposure decreases sperm quality: association between sperm parameters and urinary pesticide levels [J]. J Appl Toxicol 28(5):674–680
18. Butler-Dawson J, Galvin K, Thorne PS et al (2016) Organophosphorus pesticide exposure and neurobehavioral performance in Latino children living in an orchard community [J]. Neuro Toxicol 53:165–172
19. Abreu-Villaça Y, Levin ED (2017) Developmental neurotoxicity of succeeding generations of insecticides [J]. Environ Int 99:55–77
20. Zhang Y, Yang Z, Xie F et al (2014) Research advance of OPs neural development toxicity mechanism [J]. J Anhui Agric Sci 42(12):3561–3564
21. Timoroğlu İ, Yüzbaşıoğlu D, Ünal F et al (2014) Assessment of the genotoxic effects of organophosphorus insecticides phorate and trichlorfon in human lymphocytes [J]. Environ Toxicol 29(5):577–587
22. Soderlund DM, Clark JM, Sheets LP et al (2002) Mechanisms of pyrethroid neurotoxicity: implications for cumulative risk assessment [J]. Toxicology 171(1):3–59
23. Soderlund DM (2012) Molecular mechanisms of pyrethroid insecticide neurotoxicity: recent advances [J]. Arch Toxicol 86(2):165–181
24. Saillenfait A, Ndiaye D, Sabaté J (2015) Pyrethroids: exposure and health effects – an update [J]. Int J Hyg Environ Health 218(3):281–292
25. Laugeray A, Herzine A, Perche O et al (2017) In utero and lactational exposure to low-doses of the pyrethroid insecticide cypermethrin leads to neurodevelopmental defects in male mice—an ethological and transcriptomic study [J]. PLoS One 12(10):e184475
26. Shafer TJ, Meyer DA, Crofton KM (2005) Developmental neurotoxicity of pyrethroid insecticides: critical review and future research needs [J]. Environ Health Perspect 113(2):123–136
27. Li Q, Kobayashi M, Kawada T (2015) Carbamate pesticide-induced apoptosis in human T lymphocytes [J]. Int J Environ Res Public Health 12(4):3633–3645
28. Li Q, Kobayashi M, Kawada T (2014) Carbamate pesticide-induced apoptosis and necrosis in human natural killer cells [J]. J Biol Regul Homeost Agents 28(1):23–32
29. Rowe C, Gunier R, Bradman A et al (2016) Residential proximity to organophosphate and carbamate pesticide use during pregnancy, poverty during childhood, and cognitive functioning in 10-year-old children [J]. Environ Res 150:128–137
30. Popovska-Gorevski M, Dubocovich ML, Rajnarayanan RV (2016) Carbamate insecticides target human melatonin receptors [J]. Chem Res Toxicol 30(2):574–582

31. Soloneski S, Kujawski M, Scuto A et al (2015) Carbamates: a study on genotoxic, cytotoxic, and apoptotic effects induced in Chinese hamster ovary (CHO-K1) cells [J]. Toxicol In Vitro 29(5):834–844
32. Piel C, Pouchieu C, Migault L et al (2018) Increased risk of central nervous system tumours with carbamate insecticide use in the prospective cohort AGRICAN [J]. Int J Epidemiol 48:512–526
33. Dhouib I, Jallouli M, Annabi A et al (2016) From immunotoxicity to carcinogenicity: the effects of carbamate pesticides on the immune system [J]. Environ Sci Pollut Res 23(10):9448–9458

Flame Retardants: Exposure, Biomarkers, and Health Risks

Yuting Jin and Shangqin Chen

Abstract Flame retardants have been extensively used in various fields of production and daily life, such as aerospace, transportation, and interior decoration, to ensure fire safety. They are mainly divided into inorganic, halogenated organic, nitrogen-containing, and phosphorus-containing compounds, among which, brominated flame retardants and organophosphorous flame retardants are the most commonly used flame retardants. However, with reports toward adverse effects of flame retardants on human being gradually increasing, more and more attention has been paid to the health risks caused by the use of flame retardants. This chapter introduces the exposure of flame retardants to environment and human, as well as sensitive biomarkers of flame retardants, and emphatically analyzes the toxic effects of flame retardants on health risks including endocrine disruption, neurotoxicity, reproductive inability, immunotoxicity, hepatotoxicity, lung toxicity, carcinogenesis, preterm birth, developmental retardation, and children behavioral problems.

Keywords Flame retardants · Environment exposure · Biomarkers · Health risk

1 Introduction

Fire accidents pose immense threats to human life and property; therefore, fire safety has become an important issue in the national economy and social development. Flame retardants (FRs) have been added to flammable materials used in various fields of production and life, including aerospace, transportation, interior decoration, to ensure fire safety. The European Commission estimates that the use of FRs has reduced fire deaths in Europe by 20% over the past decade. At present, the global total amount of FRs is second only to platicizers in all kinds of plastic additives, and have a very broad prospect for development. However, the accompanying deleterious effects caused by FRs are worth considering.

Y. Jin · S. Chen (✉)
The Second Affiliated Hospital and Yuying Children's Hospital of Wenzhou Medical University, Wenzhou, China

Y. Zhang (ed.), *Emerging Chemicals and Human Health*,
https://doi.org/10.1007/978-981-32-9535-3_8

FRs are divided in four major categories depending on their chemical composition: inorganic, halogenated organic, nitrogen-containing, and phosphorus-containing compounds. Brominated flame retardants (BFRs) and organophosphate flame retardants (OPFRs) are the most commonly used FRs. Annual global consumption of FRs is about 2,000,000 tons, of which 35% are organophosphorus and brominated, and 85% are used in textile and rubber products [1].

BFRs are an important group of flame retardants, and their production and use have more than 30 years of history. It is recognized as a highly efficient flame retardant, widely used in electronics, electrical, chemical, transportation, building materials, textiles, petroleum, mining, and other fields. There are more than 70 varieties of BFRs, including polybrominated biphenyl ethers (PBDEs), polybrominated biphenyls (PBBs), tetrabromobisphenol A (TBBPA), and hexabromocyclododecane (HBCD), among which PBDEs are one of the most widely used BFRs. However, BFRs have the characteristics of persistence, bioaccumulation, and ecological toxicity, as well as neurological, immune, reproductive toxicity and endocrine interference. Therefore, in 2003, the EU introduced two bans, RoHS (Restriction of Hazardous Substances Directive) and WEEE (Directive on Waste Electrical and Electronic Equipment). As of July 1, 2006, all electronic and electrical equipment (EEE) newly introduced into the market in EU countries cannot contain PBDEs. And in 2009, the Stockholm Convention listed of PBDEs as a persistent organic pollutant and restricted the production and use. Subsequently, a number of countries and regions, including the EU, the United States, and Canada, issued a series of documents to restrict or prohibit the use of PBDEs. Chlorinated flame retardants (CFRs) are another group of BFRs, including polychlorinated biphenyls (PCBs) and dechlorane Plus (DP). The chemical structure of PCBs is similar to that of PBDEs.

The restrictions and ban of classical FRs have led to the use of alternatives like OPFRs and novel brominated flame retardants (NBFRs) which are now classified also as emerging pollutants. OPFRs, one of the most widely used groups of FRs, mainly include organophosphorus esters (OPEs), tris(1,3- dichloro-2-propyl) phosphate (TDCIPP), tris(chloropropyl) phosphate (TCIPP), tris(2-chloroethyl) phosphate (TCEP), and triphenyl phosphate (TPHP). Compared with BFRs, OPFRs have the advantages of high flame retardancy, less corrosion of decomposition products, and less toxic substances. In addition, OPFRs not only have flame-retardant properties but also have plasticizing, thermal stability and other functions, which play a very important role in improving the comprehensive properties of polymer materials. However, in recent years, as the output and consumption of OPFRs have been increasing year by year, the environmental effects of OPFRs have also aroused widespread concern.

Other commonly used organic flame retardants include iremaster 550 (FM 550), perfluorooctane sulfonates (PFOS), and perfluorooctane sulfonyl fluoride (PFOSF). FM 550, a Chemtura mixture of brominated and nonhalogen flame retardants used in foam-based products, significantly reduces the risk of fires from flexible polyurethane foam used as cushioning in furniture and other products and is recently identified as a common contaminant in household dust.

2 Environmental Exposure to FRs

FRs have been used for decades, and thus, they have been detected in almost all environmental compartments. FRs are transported through the environment and either deposited in sediments or are bioaccumulated. Figure 1 depicts how a flame retardant is transported in the environment. The figure shows many mechanisms responsible for the transport and fate of FRs. Various sources can be responsible for FRs entering into water environment, which may include FRs for the production of industrial wastewater, flame-retardant polymers, and plastic products, such as facilities recycling plastic, incinerators, sewage treatment plants, electrical appliances, and metals for recovery from electronic equipment and accidental fires, sewage discharge, and to a lesser extent by atmospheric deposition.

There is also evidence of various biota transporting contaminants, which is known as "biotransport." A lot of studies have proved that salmon is involved in the biotransport of POPs including PBDEs, and have contaminated lake food webs. This phenomenon is usually observed for fishes that undergo migration for their spawning period. Once entering into the aquatic system FRs can undergo fractionation, accumulation, or lost to the atmosphere. Sediments occur mainly in lakes, river sediments, and aquatic organisms. FRs are mostly lipophilic and have a high binding affinity for suspended particles, which is easy to accumulate in sediments.

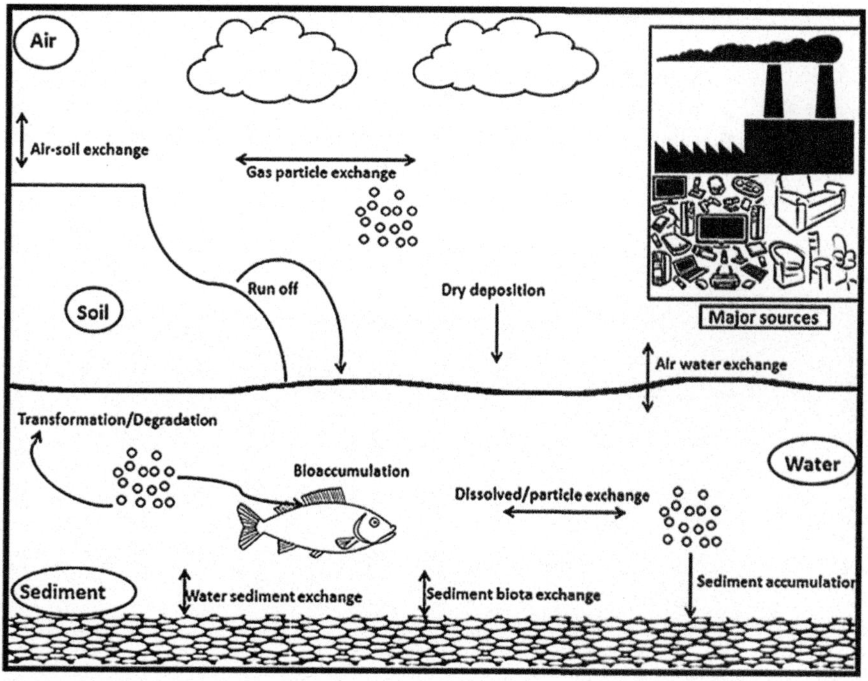

Fig. 1 Transport and fate mechanisms of flame retardants in the freshwater ecosystem [1]

Suspended particles are considered as important carriers for higher BDEs in aquatic environment. FR exposures of water, sediments, air, diet, and human are detailed below.

2.1 FR Exposure in Water Environment

FRs with high hydrophobicity tend to associate with particulars in water, with only a very small fraction dissolved in it (freely dissolved fraction). Filtering eliminates suspended and larger particles and makes water samples more representative of the dissolved fraction of FRs which is also bioavailable for aquatic biota. There have been many studies on FR detection in river and lake waters globally, but still several countries and regions have no inventories of different groups of FR concentrations in riverine ecosystems. Due to the low solubility of FRs in water, studies on river waters are very scarce.

The FRs in the water environment are mainly caused by the leakage and volatilizing of FR products. Currently, it is widely believed that sewage discharge is the main source of FRs in surface water. OPFRs reported in several studies exist in river ecosystems due to anthropogenic pressures and waste disposal facilities. TCEP concentrations as measured in different rivers were 217.5 ng/L in River Aire, 85 ng/L in Spain, 71.5 ng/L in Germany, and 71.5 ng/L in Austria [2–4]. The FR pollution levels are in industrially affected rivers. For example, the OPFR concentration in a river in the United Kingdom from urban and industrial discharge was 75–26,050 ng/L. Among them, the maximum concentration of TCIPP in surface water was 26 ng/L, higher than that in other rivers. OPFR concentrations reported here were higher than those of PBDEs. Temporal trend of OPFRs suggested that the concentrations varied greatly over time (Fig. 2).

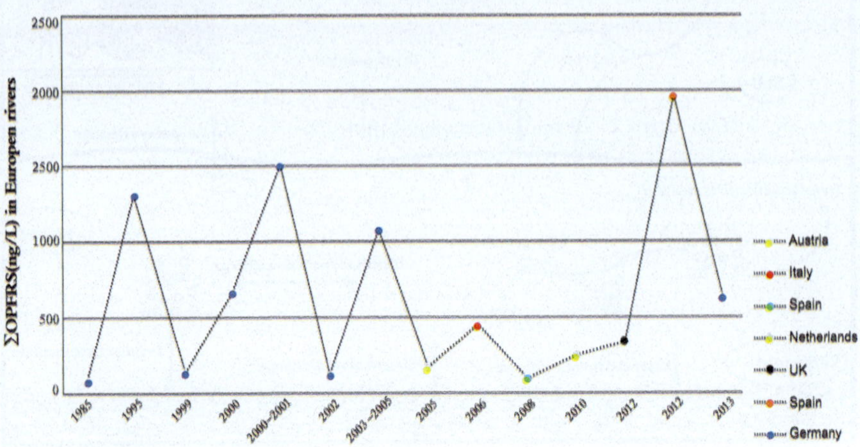

Fig. 2 Concentration of ΣOPFRs in European rivers by year and country of collection [1]

In addition, individual sites with the highest PBDE measurements were found in shellfish taken from Anaheim Bay, California, and four sites in the Hudson Raritan Estuary in the United States. High PBDE concentrations of PBDEs have also been found in the watersheds that include the Southern California Bight, Puget Sound, the central and eastern Gulf of Mexico off the coast of Tampa and St. Petersburg, and the waters of Lake Michigan near Chicago and Indiana. According to previous studies, solubility of PBDEs decreased with the increase of bromine content, while log K_{ow} increased with the increase of bromine content, which proves that BDE-47 and BDE-99 occupy a high proportion in waters all over the world. A study at River Aire showed that only BDE-209 was present in water samples among all PBDEs, with concentrations ranging from 17 to 295 ng/L.

In Denmark, ΣHBCD in aquatic systems ranged from 0.052 to 2.9 ng/L with α-HBCD accounting for 40–60% of ΣHBCD, which may be due to its higher solubility (48 μg/L) than β-HBCD and γ-HBCD. A recent study in River Chenab (Pakistan) showed high concentrations of ΣPBDE which ranged from 0.48 to 73.4 ng/L, much higher than the concentrations of ΣDP from 0.01 ng/L to 4.58 ng/L, reflecting the broad use of the technical BDE formulations in this area.

2.2 *Accumulation of FRs in Sediment*

Sediments are considered to be the ultimate sink of organic pollutants with high hydrophobicity and hydrophilicity. FRs either accumulates in biota or is absorbed into sediments after entering aquatic systems. Various studies have shown that FRs tends to accumulate in sediments representing contaminated and industrialized sampling areas.

Sediment record and atmospheric deposition of BFRs was compared with legacy POPs in Lake Thun, Switzerland, and it was revealed that PBDEs and HBCDs started depositing since the 1980s from dated sediment cores [5]. While in more recent sediment layers, BFRs still had an increasing trend similar to most of the OPFRs. Sediments levels analyzed from the origin of rivers were low but increased significantly as the river approached its mouth. Rivers and lakes affected by industrial and urban areas are more likely to accumulate FRs in sediments [6].

A study showed that inhabitants living within 10 km, especially those living within 2 km of any solid-waste facility, had higher serum levels of PBDEs than those living >10 km away, suggesting that solid-waste disposal sites may be an important source of PBDE exposures to the general population living in close proximity to such sites [7]. Researchers had studied FRs in sediments from the River Po and the River Lambro in Italy. They found that the lowest concentration determination of TBBPA was 1.2 ng/g dw, while ΣPBDEs showed a concentration up to 3132.2 ng/g, proved to the dominant in BDE-209 [8].

Decabromodiphenyl ethane (DBDPE) and DP are well studied in emerging FRs. DBDPE with concentrations ranging from 91 ng/g to 435 ng/g was detected in sediment samples in River Besòs in Spain, and it was also detected at the site with the highest concentration of BDE-209, indicating that BDE-209 was replaced by DBDPE.

2.3 FRs Exposure in Ambient and Indoor Air

Air is an important medium for the transmission and dissemination of FRs, and the air exposure of FRs is most closely related to human health. Monitoring of the atmosphere over the Great lakes showed that the total concentration of OPFRs was 120–2100 pg/m^3. Moreover, studies have found that the exposure to indoor PBDEs is far more serious than that of outdoors, suggesting the extensive application of PBDEs in furniture and building materials.

Flame-retarded products have a direct chain of emissions from which outgas PBDEs to indoor air and increase the concentrations indoors. The indoor air is barely changed through the ventilation systems and released to the outdoors. Once released, indoor air is diluted into larger volumes of outdoor air, which is then transported from buildings and urban areas, leading to the gradient from urban to rural areas seen in other studies. In the atmosphere, PBDEs including BDE-209 can also carry out long-distance atmospheric transmission. Therefore, PBDEs released from indoor applications are sources of pollution in the outdoor environment, leading to food web pollution, which in turn leads to human food source pollution.

Zhang et al. [9] recorded concentrations of PBDEs in indoor air and dust at 20 indoor sites (homes, offices, and laboratories) in Toronto, Canada, in 2006. Despite the little difference, the average concentration of BDE-7 in the air of offices (0.79 ng/m^3) exceeded that of homes (0.49 ng/m^3). The concentration of BDE-7 in this indoor air far exceeded that in outdoor air in Toronto, indicating that PBDEs in indoor air are a source of PBDEs in outdoor air. It is evident and logical that a room's contents and building materials are reflected in its air and dust concentrations and compound profiles. In general, offices have higher PBDE and PCB concentrations than homes. Even 27 years after construction, offices had more than 30 times higher PCB air concentration than outdoors. It is presumed that PCBs are incorporated in the building's construction materials and PBDEs are emitted from the electronic equipment first, followed by carpet and possibly PUF furniture that dated to just prior the 2004 restrictions on PBDE production.

Notably, PBDEs have the ability to achieve global distribution. Because even in places far away from people, such as the sparsely populated Qinghai-Tibet Plateau and the Antarctic, there is also the distribution of PBDEs.

2.4 FRs Exposure via Diet

In addition to water, sediments, and air, diet is also one of the main pathways of human exposure to PBDEs. PBDEs can bioaccumulate and amplify in the food chain, which is particularly prominent in the aquatic ecosystem. Due to its high lipid solubility and strong ability to combine with particulate matters, except for a small part of dissolved PBDEs, most PBDEs can accumulate in sediments. Although their concentration in water is extremely low, once absorbed by the organism, they can accumulate in organisms, amplify through the food chain, and result in human exposure.

For example, different concentrations of PBDEs were detected in pork, chicken, vegetables, fruits, eggs, oils, and other foods in Japan, Sweden, and Spain. Among them, the concentrations of PBDEs were higher in oil, meat products, and eggs and lower in fruits and vegetables. In Shenzhen of China, fatty fish and shellfish were the main sources of adult consumption of PBDEs. But other studies have shown that dietary exposure becomes an important route of exposure only when the airborne PBDE levels are relatively low.

3 Exposure Assessment of FRs in Humans

Researchers are characterizing human exposure to FRs based on biomonitoring studies of FRs (e.g., blood, breast milk, placenta, and hair), with particular focus on external exposure routes (e.g., diet and dust ingestion, inhalation of air) and internal doses [10]. Although the contribution of different exposure pathways to human exposure to FRs varies greatly among different compounds, samples, and populations in different regions, people living in North America tend to have serum and breast milk PBDE levels that are one to two orders of magnitude higher than those in Europe and Asia [11]. However, FR exposure through food intake is relatively low, and the intake through dust (especially indoor dust) contributes to its total intake. It is estimated that dust/soil ingestion and skin contact can explain 90% of the daily exposure of adults to PBDEs. Levels of PBDEs in the human body have increased more than 100-fold in about 25 years of human exposure. With the production, use, and destruction of the FR-related products, the environmental exposure level of FRs gradually increases, as well as the increasing human health risks. The human exposure pathways of FRs are shown in Fig. 3.

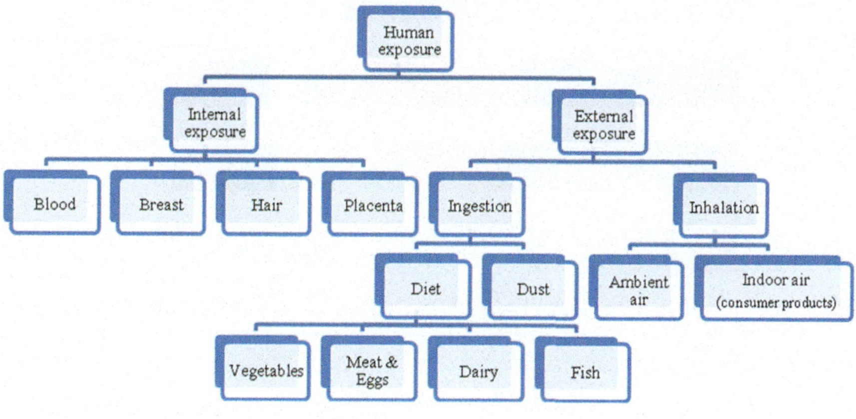

Fig. 3 Human exposure pathways of FRs

The enrichment of PBDEs in biological larva is relatively high, such as Chinese sturgeon eggs [12]. However, the contents of PBDEs in mature organisms including polar bears and male minks are relatively low [13, 14]. Besides, PBDEs are mainly concentrated in human blood. There is, of course, some accumulation of PBDEs in umbilical cord blood and feces at birth and breast milk of lactating women [15–17]. Breast milk is the main source of infants to intake PBDEs. A study on the content of PBDEs in breast milk and infant formula from China found that the concentrations of PBDE decreased from 14.8 to 4.85 pmol/g lipid weight between 2006 and 2012, with a rate of decrease by about half every 4 years [18]. PBDEs in breast milk were much higher than those in infant formula, indicating that children who are breastfed have a higher exposure to PBDEs. Currently, the overall level of PBDEs in breast milk is relatively low in Europe and Asia but highest in North America. However, with the increasing scale of electronics processing and recycling in Asia, the concentration of PBDEs in human breast milk of the Asian population has a tendency to increase further.

The level of PBDEs in umbilical cord blood is an important indicator for monitoring the exposure of PBDEs. The serum of umbilical cord blood contains PBDEs, indicating that PBDEs could pass through the placental barrier and further threaten the growth and development of fetus. Previous studies have shown that the concentration of PBDEs in maternal serum and umbilical cord blood serum is closely related to the concentration of PBDES in maternal serum and placenta. In Shin's [19] study, 41 mother-neonate pairs were recruited and provided samples of maternal serum, umbilical cord serum, breast milk, and house dust. As shown in Fig. 4, the most abundant congener in house dust was BDE-209. Although BDE-47 was most abundant in maternal serum, attributed to its relatively long half-life (1.8 years) in humans, followed by BDE-153, the congener composition was notably different

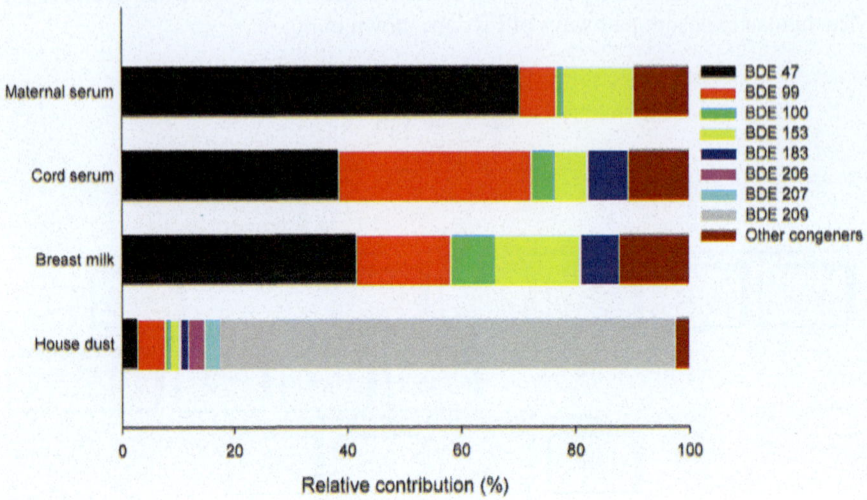

Fig. 4 Relative contributions of PBDE congeners in human samples and house dust [19]

according to the biological matrix. The levels and detection rates of BDE-99 in umbilical cord serum were increased, and the ratio of BDE-47 to BDE-99 in this matrix was about 1:1, which strongly suggested that the placenta was not an effective barrier to limit the transport of PBDEs. In breast milk, the most abundant congener was BDE 47, followed by BDE-99 and BDE-153. Significant correlations were found between BDE-209 in dust and BDE-47 and BDE-99 in umbilical cord serum, suggesting considerable dosage of lower-brominated BDE in mother-neonate pairs might come from their house dust.

It is now well documented that exposure early in embryonic development to commonly used chemicals alters gene expression patterns that can lead to altered health later in life. Early-life exposure to chemicals with endocrine disruption potential alters gene expression profiles that are associated with altered morphology and physiology, such as impaired fertility and reproductive tract development, behavioral changes, altered metabolism, and obesity. Parental genomes are based on genomes inherited from parents, including cumulative genetic mutations, and epigenetic modifications that occur before the new offspring are fertilized. Environmental factors have been shown to modify the parental genome, so that the genetic makeup of any offspring is a combination of the parental genetic genome and environmental influences on the germline during maturation. Environmental factors (e.g., diet, stress, and pollutants) can also modify the genomes of developing embryos through classic selection and mutation or through epigenetic mechanisms at the somatic and germline levels. These changes may produce health and disease predispositions in an individual's current lifetime. Future intergenerational effects can also be determined by modifying the germline genome or epigenome after exposure to the virus in an individual's lifetime [20].

4 Biomarkers of FRs

FRs can be detected in a variety of tissues, such as hair, breast milk, blood, placenta, and urine. However, urinary FRs, as well as its metabolites, appear to be the preferred noninvasive biomarkers for identifying and quantifying human exposure to FRs. Urinary FRs as well as its metabolites provide comprehensive information about the total human burden, covering all types of sources and exposure pathways (including inhalation, skin absorption, and oral absorption), so they can be used to quantify an individual"s exposure [21]. The information about the metabolism of FRs in humans is still limited, and there are differences among different compounds.

The general chemical formula of PBDEs is $C_{12}H_{(0-9)}Br_{(1-10)}O$. Figure 5 shows the chemical structure of PBDEs. PBDEs are divided into 10 homologous groups with a total of 209 homologous species, according to the number and position of bromine atom substitutions, following the International Union of Pure and Applied Chemistry (IUPAC) nomenclature. Table 1 details the congener distributions of PBDEs. However, there are only dozens of PBDE congeners which are common in the envi-

Fig. 5 The structural of PBDEs

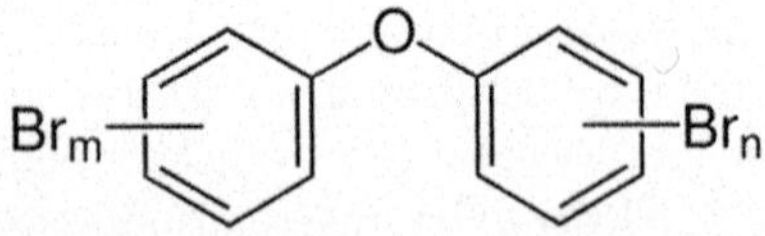

Table 1 Nomenclature and congener distributions of PBDEs

Homologous group	Molecular formula	Molecular weight	Number	Homologous series
Mono-PBDEs	$C_{12}H_9OBr$	249.0	3	BDE-1, -2, **-3**
Di-BDEs	$C_{12}H_8OBr_2$	327.9	12	BDE-4, -5, -6, -7, -8, -9, -10, -11, -12, -13, -14, -15
Tri-BDEs	$C_{12}H_7OBr_3$	406.9	24	BDE-16, **-17**, -18, -19, -20, -21, -22, -23, -24, -25, -26, -27, **-28**, -29, -30, -31, -32, **-33**, -34, -35, -36, -37, -38, -39
Tetra-BDEs	$C_{12}H_6OBr_4$	485.8	42	BDE-40, -41, -42, -43, -44, -45, -46, **-47**, -48, -49, -50, -51, -52, -53, -54, -55, -56, -57, -58, -59, -60, -61, -62, -63, -64, -65, **-66**, -67, -68, -69, -70, -71, -72, -73, -74, -75, -76, -77, -78 -79, -80, -81
Penta-BDEs	$C_{12}H_5OBr_5$	564.7	46	BDE-82, -83, -84, **-85**, -86, -87, -88, -89, -90, -91, -92, -93, -94, -95, -96, -97, -98, **-99**, **-100**, -101, -102, -103, -104, -105, -106, -107, -108, -109, -110, -111, -112, -113, -114, -115, -116, -117, -118, -119, -120, -121, -122, -123, -124, -125, -126, -127
Hexa-BDEs	$C_{12}H_4OBr_6$	643.6	42	BDE-128, -129, -130, -131, -132, -133, -134, -135, -136, -137, **-138**, -139, -140, -141, -142, -143, -144, -145, -146, -147, -148, -149, -150, 151, -152, **-153**, **-154,** -155, -156, -157, -158, -159, -160, -161, -162, -163, -164, -165, -166, -167, -168, -169
Hepta-BDEs	$C_{12}H_3OBr_7$	722.5	24	BDE-170, -171, -172, -173, -174, -175, -176, -177, -178, -179, -180, -181, -182, **-183**, -184, -185, -186, -187, -188, -189, **-190,** -191, -192, -193
Octa-BDEs	$C_{12}H_2OBr_8$	801.4	12	BDE-194, -195, **-196**, **-197,** -198, -199, -200, -201, -202, **-203**, **-204**, **-205**
Nona-BDEs	$C_{12}HOBr_9$	880.3	3	**BDE-206, -207, -208**
Deca-BDEs	$C_{12}OBr_{10}$	959.2	1	**BDE-209**

[a]Bold type indicates common PBDEs congeners

ronment, such as BDE-17, BDE-28, BDE-33, BDE-47, BDE-66, BDE-99, BDE-100, BDE-138, BDE-153, BDE-154, BDE-183, BDE-190, BDE-196, BDE-203, BDE-204, BDE-205, BDE-206, BDE-207, BDE-208, and BDE-209. In addition, the dominant metabolites of PBDEs are hydroxylated BDE (OH-PBDEs) and methoxylated BDE (MeO-PBDEs).

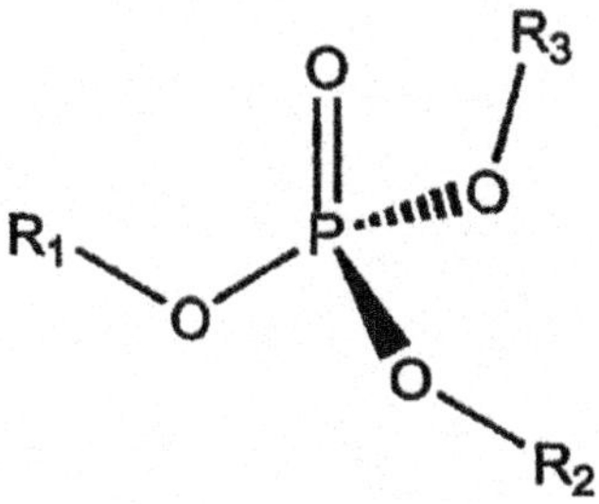

Fig. 6 The structure of OPFRs

Figure 6 shows the common molecular structure of OPFRs. According to the different substituent, OPFRs can be divided into alkyl phosphate, aryl phosphate, and halogenated alkyl phosphate. Triphenyl phosphate (TPHP), 2-ethylhexyl diphenyl phosphate (EHDPP), and tricresyl phosphate (TCrP) are biomarkers of OPRFs, and their main metabolites are 2-ethyl-5-hydroxyhexyl diphenyl phosphate (5-OH-EHDPP), 4-hydroxyphenyl diphenyl phosphate (4-OH-TPHP), and 3-hydroxy-4-methylphenyl di-*p*-tolyl phosphate (3-OH-MDTP) [22]. Table 2 details the parent compounds and metabolites of OPFRs.

FM550 consists of 2-ethylhexyl-2,3,4,5-tetrabromobenzoate (TBB), di(2-ethylhexyl) tetrabromophthalate (TBPH), triphenyl phosphate (TPP), and isopropylated triphenylphosphate isomers (ITPs), with the structures shown in Fig. 7. Urinary tetrabromobenzoic acid (TBBA) is a biomarker of FM 550 [23].

5 FRs and Health Risks

The environmental behavior of flame retardants has been studied since the 1990s. Some individual substances have been found to have harmful effects on the human body.

5.1 Endocrine Disruption

BFRs are endocrine-disrupting chemicals (EDCs) that can reduce peripheral thyroid hormone (TH) levels. There are various mechanisms of thyroid disruption and different chemicals have different levels of interference on the hypothalamic-pituitary-thyroid axis. Mechanisms of action may involve the sodium-iodide symporter, thyroid peroxidase enzyme, receptors for THs or thyroid stimulating hormone (TSH), transport proteins, or cellular uptake mechanisms. Through effects on iodothyronine deiodinases or hepatic enzymes, the peripheral metabolism of the THs can be affected. In the rats exposed to PBDEs, serum thyroid hormone was significantly reduced, with the broken homeostasis of thyroid hormone [24]. The disruption of thyroid hormone homeostasis may be caused by the competitive binding of

Table 2 Parent compounds and metabolites of OPFRs [21]

Parent OPFR		OPFR metabolite	
Full name (CAS number)	Abbreviation	Full name	Abbreviation
Tris(2-chloroethyl) phosphate (115–96-8)	TCEP	Bis(2-chloroethyl) phosphate	BCEP
Tris(1-chloro-2-propyl) phosphate (13674–84-5)	TCIPP	Bis(1-chloro-2-propyl) phosphate Bis(1-chloro-2-propyl) 1-hydroxy-2-propyl phosphate	BCIPP BCIPHPP
Tris(1,3-dichloro-2-propyl) phosphate (or isopropyl) (13674–87-8)	TDCIPP	Bis(1,3-dichloro-2-propyl) phosphate	BDCIPP
Tri-*n*-butyl phosphate (126–73-8)	TNBP	Di-*n*-butyl phosphate	DNBP
Tris(2-ethylhexyl) phosphate (78–42-2)	TEHP	Bis(2-ethylhexyl) phosphate	BEHP
Mono-substituted isopropyl triphenyl phosphate (isopropylphenyl diphenyl phosphate) (several isomers: 55864–04-5, 69,515–46-4, 64,532–94-1)	Mono-ITP	Isopropylphenyl phenyl phosphate	Ip-PPP
Tris(2-butoxyethyl) phosphate (78–51-3)	TBOEP	Bis(2-butoxyethyl) phosphate Bis(2-butoxyethyl)-(2-hydroxyethyl) phosphate	BBOEP BBOEHEP
Triphenyl phosphate (115–86-6)	TPHP	Diphenyl phosphate	DPHP
Tricresyl phosphate (1330-78-5) *Ortho*, *meta*, and *para* isomers (78–30-8, 563–04-2, 78–32-0, respectively)	TCP ToCP, TmCP, TpCP	Dicresyl phosphate	DCP
2-ethylhexyl diphenyl phosphate (1241-94-7)	EHDPP	5-hydroxy-2-ethylhexyl diphenyl phosphate	5-OH-EHDPHP

thyroid transporter proteins, such as thyroxin-binding globulin and thyroxine transporter, thus accelerating the excretion and metabolism of thyroid hormone, changing the activity of thyroid hormone receptors, and interfering with the activity of deiodinase [25]. It was reported that hydroxylated and methoxylated metabolites (OH-BDEs and MeO-BDEs) of PBDEs had similar biological effects as PBDEs, and OH-BDEs had higher binding affinity for human transthyretin (TTR) than PBDEs [26].

During gestation development, changes in maternal thyroid hormone are very important, because the fetus does not start to synthesize thyroid hormone until the 18th–22nd week of gestation, but after that time, the fetus cannot produce enough thyroid hormones, so it still relies on the maternal input to control the stability of thyroid hormone in the fetus [27]. Therefore, even small changes in thyroid homeo-

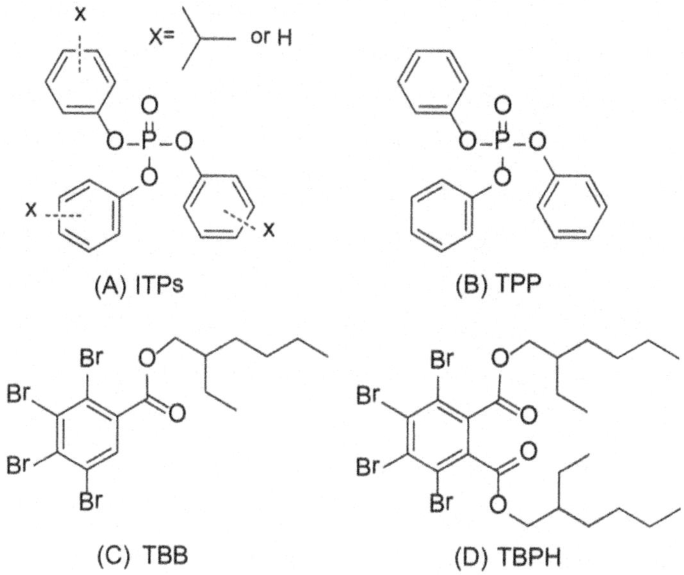

Fig. 7 Chemical structures of FM550

stasis may adversely affect human health, and especially fetal neurological development may be vulnerable [28].

Studies demonstrate that like PBDEs, TBBPA exposure in animals leads to decreases in circulating T4 levels. TBBPA has also been shown to inhibit thyroid deiodinase type 1 activity and strongly bind to and inhibit estrogen and thyroid sulfotransferases [29–31]. FM550 exposure across gestational and lactational stages resulted in significant increases in dam serum thyroxine levels as well as early puberty and metabolic dysfunction in offspring [32]. Furthermore, a metabolite of one of the BFRs in FM550 was shown to significantly decrease serum T3 levels in rats and activate a nuclear receptor involved in adipogenesis [33].

FRs as EDCs also lead to obesity, which is not only a lifelong overweight problem, but has been identified as a potential risk factor for many diseases (e.g., metabolic syndrome, diabetes, cardiovascular disease, and cancer) [34–36]. Coincidentally, the development of these diseases is associated with exposure to EDCs. In this context, it is worth noting that many EDCs are lipophilic, and POPs in particular are known to bioaccumulate in body fat over the years. Thus, the association between obesity and disease seems to be related not only to the fat deposition but also to the larger volume of fat and the larger retention of lipophilic pollutants. Figure 8 shows the potential for a "vicious spiral" [37] to be set up whereby obesogens act to increase the amount of fat stored. And a more conservative version of the lipophilic obesity gene followed. This will cause more spirals and more lipophilic contaminants to run aground in the future. In this sense, the obesity gene may be self-fulfilling and increase its ability to retain.

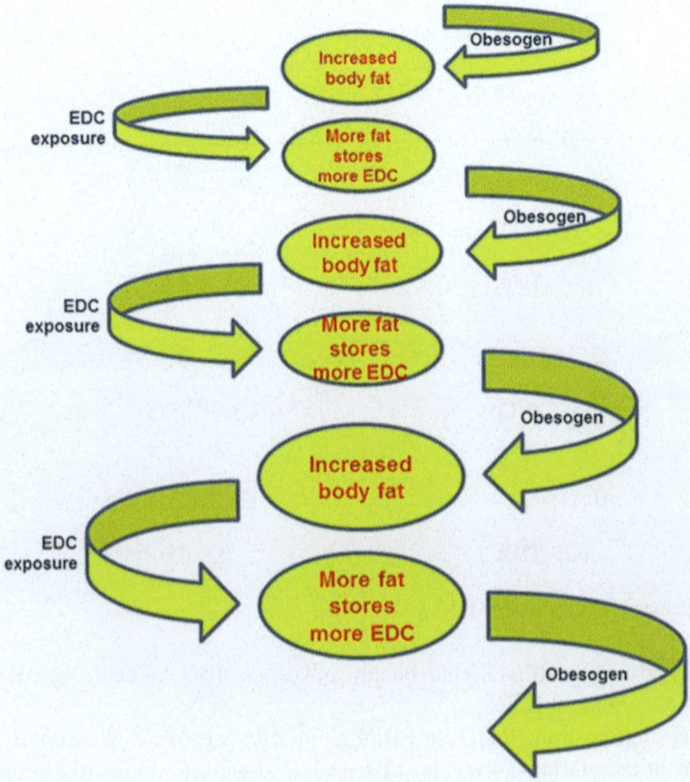

Fig. 8 The "vicious spiral" of obesogenic activity and lipophilic properties of endocrine disruptors [37]

5.2 *Neurotoxicity*

The neurobehavioral toxicity of FRs may affect the normal growth and development of the nervous system at the early growing stages of organisms, and may cause adverse effects on their healthy development and viability. OPFRs, with a similar structure to organophosphorus pesticides, have been proved to be a neurobehavioral teratogen. In addition, in terms of central nervous system, neural behavioral injury induced by PBDE exposure may firstly reflect the changes of neuron connectivity and neuroplasticity during brain development. For example, by inhibiting axonal growth of primary and secondary motor neurons during the early developmental stages, BDE-47 exposure significantly affected spontaneous movement and decreased touch response and free-swimming speed in developing zebrafish [38].

Besides, PBDEs may indirectly affect the developing brain by altering the homeostasis of thyroid hormones. THs regulate the cell structure, neuron growth, and synapse formation of neurons, and their receptors are widely distributed in the central nervous system. Any lack or increase in thyroid function (hypothyroidism or hyperthyroidism) can result in irreversible damage, morphological and cellular abnormalities, developmental delays, tissue disorders, and physical retardation. These include abnormal neuronal proliferation, migration, decreased dendritic densities, and dendritic arborizations. This drastic effect may result in the loss of important functions of neurons, which may lead to biochemical dysfunction. This could explain the physiological and behavioral changes observed in animals or humans during thyroid dysfunction. Moreover, the sensitivity of TH is manifested in the neonatal period, as well as the prebirth period. Changes in thyroid hormone during development may result in brain injury, if not corrected soon after birth [27].

5.3 Reproductive Toxicity

Animal studies have shown that PBDEs interfere with estrogen- and androgen-mediated processes. BDE-47 exposure can decrease sperm motility and sperm energy rate, and cause reproductive inability. An et al. found that 4-day exposure to BDE-71 with a daily concentration of 60 mg/kg or 5-day exposure to BDE-71 with a daily concentration of 120 mg/kg led to changes in the secretion of sex hormones in rats and the occurrence of puberty delay [39]. In addition, PBDEs can also damage the mitochondria in the ovaries of adult female mice, affect lipid synthesis, and reduce the number of sperm cells and sperm in male offspring [40]. The concentration of BDE-153 in serum of young men was negatively correlated with their sperm motility [41]. Main et al. tested 86 samples of breast milk and showed that the concentration of PBDEs in breast milk was significantly higher in boys with cryptorchidism than in controls [42]. Chao et al. found PBDEs appeared to have potential to prolong length of average menstrual cycle and delay the age when menstruation periods begin coming regularly [43].

Besides, synthetic flame retardant, tris(1,3-dichloro-2-propyl) phosphate (TDCIPP), inhibited the sperm development of male zebrafish, with decreasing the diameter of the eggs of F1 generation and increasing the malformation rate. Li et al. also found that exposure to environmentally relevant concentrations of TDCIPP significantly decreased fecundity of F0 and F1 generations [44].

5.4 Immunotoxicity

The nervous system, endocrine system, and immune system in the organism have complex mechanisms, which affect and regulate each other, and jointly promote the growth and development of the organism and the regulation of physiological functions. When BFRs cause neurotoxicity and thyroid toxicity, the immune system will also be affected. For example, BFRs affected the structure of spleen and thymus,

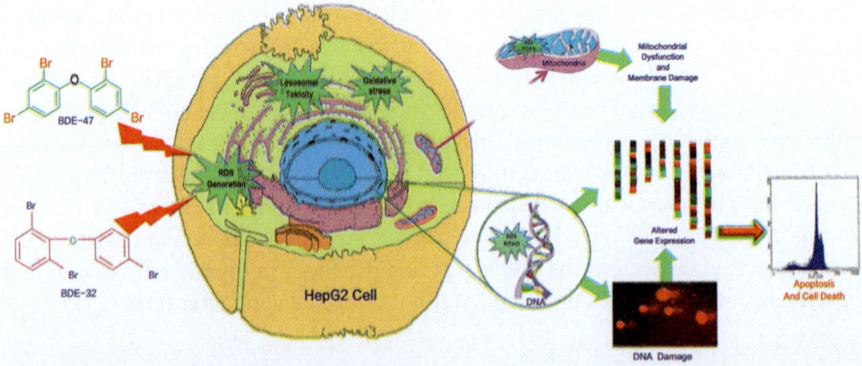

Fig. 9 BDE-47 and BDE-32 exposures trigger apoptosis in HepG2 cells through DNA damage and mitochondrial dysfunction [47]

inhibited the sheep red blood cell response, reduced the weight of thymus, and decreased the number of spleen cells. BDE-47 reduced the level of inflammatory cytokines in children but enhanced the immune response of peripheral blood monocytes in children with autism spectrum disorder [45]. Besides, the chicks exposed to PBDEs had a strong phytohemagglutinin response, which changed the structure of the spleen and thymus, thus damaging the immune system [46].

5.5 *Hepatotoxicity*

The hepatic toxicity of FRs is manifested as liver microsomal enzyme activity induction, hepatomegaly, degenerative histopathological changes, and liver cancer. Previous studies have found thyroid hyperplasia, liver volume increase, and hyaline degeneration of the kidney in mice exposed to BDE-209 virus, indicating an inhibitory effect on liver function. In addition, it was also found that feeding rats with PBDEs for 13 weeks resulted in significantly increased porphyrin content in the liver and urine of rats.

Saquibet al [47]. evaluated the in vitro low-dose hepatotoxic effects of two flame retardants (BDE-47 and BDE-32) in HepG2 cells. Both exposures triggered apoptosis in HepG2 cells through DNA damage and mitochondrial dysfunction (Fig. 9). BDE-47 exposure resulted in the appearance of subG1 apoptotic peak (30.1%) at 100 nmol/L, while BDE-32 arrested the cells in G2/M phase. Among the set of 84 genes, BDE-47 induced downregulation of majority of mRNA transcripts, while BDE-32 showed differential expression of transcripts in HepG2. The ultrastructural analysis revealed mitochondrial swelling and degeneration of cristae in BDE-47 and BDE-32 treated cells. Overall the data demonstrated the hepatotoxic potential of both congeners via alteration of vital cellular pathways.

5.6 Lung Toxicity

FRs can affect the respiratory system, but the mechanism of respiratory toxicity of FRs is rarely studied. Some studies have shown that penta-BDE (DE-71) and octa-BDE (DE-79) significantly increase the expression of intercellular cell adhesion molecule-1 (ICAM-1), interleukin-6 (IL-6), and interleukin-8 (IL-8) in human bronchial and pulmonary epithelial cells. FRs stimulate the respiratory system and immune system to cause inflammatory reactions, thus leading to the occurrence of asthma and other respiratory diseases. It is now thought that an important feature of the inflammatory process is the adhesion of white blood cells across vascular endothelial cells to inflammatory sites. ICAM-1 plays a role in promoting the adhesion of white blood cells and the translocation of them across endothelial cells to inflammatory sites. High levels of IL-8 have chemotaxis and activation effects on neutrophils, and can also induce T lymphocytes and eosinophils to gather in the airway, and promote eosinophils to release eosinophilic cationic proteins. Besides, neutrophils and eosinophils clustered in the airway secrete IL-8, which forms a circulation pathway, leading to and exacerbating asthma attacks.

In addition, there is a potential association between the exposure to the OPFRs in residential indoor dust and the development of asthma. A study conducted in 2014 by Araki et al. in Japan assessed this relationship and found that the tributyl phosphate was associated with the development of allergic rhinitis and asthma with an odds ratio of 2.55 and 2.85, respectively [48].

5.7 Carcinogenesis

FRs can lead to tumors including breast, pancreatic, and thyroid cancers. TDCIPP is listed as a human carcinogen under California's Proposition 65 in 2011 [49]. Subsequently, TCEP and TDBPP are listed as carcinogenic substances under California's proposition 65 in 2017. In Europe, TCEP, TDCIPP, and tributyl phosphate (TNBP) have been classified as substances suspected of causing cancer. Carcinogenicity studies, only performed for deca-BDEs, showed some effects at very high levels [50]. Hardell et al. analyzed the concentration of organic pollutants in adipose tissues of patients with pancreatic cancer and found that the concentration of PBDEs increased with time [51]. In addition, Li et al. showed that BDE-209 had the ability to regulate the proliferation and apoptosis of human breast, ovary, and uterine tumor cells [52]. Studies on patients in Swedish hospitals showed that the concentration of BDE-47 in adipose tissues was also associated with the development of non-Hodgkin's lymphoma [53].

In addition, two ClOPFRs, tris(1,3-dichloro-2-propyl)phosphate (TDCPP or TDCIPP) and tris(2-chloroethyl)phosphate (TCEP), have been designated as carcinogens, and the European Union and a number of US states have taken regulatory actions to restrict their use in certain product categories [54].

5.8 Preterm Birth

Increasing studies have evidenced that exposure to PBDEs may have adverse effects on pregnancy outcomes, such as miscarriage and premature delivery. FRs can enter the fetus through the placental barrier, thus affecting the growth and development of the fetus [55]. Zhao et al. observed, by comparing the concentration ratios between the paired samples, significantly higher fetus/blood and fetus/placenta ratios for BDE-28, BDE-99, and BDE-47 than for BDE-197, BDE-209, and BDE-153, while opposite results were found in placenta/blood ratios. Their results indicated that PBDEs entered the fetus during the first trimester and low-brominated congeners crossed the placenta more easily than high-brominated congeners, which tended to remain in the placenta [56]. In a cohort study of preterm births, more than 95% of maternal blood samples had PBDEs detected, and women with high concentrations of BDE-47 in plasma were more likely to have preterm births than women with low concentrations. Results became even more significant after adjustment for maternal race, age, and marital status. These findings suggested that high levels of maternal exposure to PBDEs might increase the risk for preterm birth [57].

However, the mechanism by which FRs cause preterm birth still remains unclear. Behnia et al. used an in vitro model of human fetal membranes to examine the reactive oxygen species (ROS) induced by PBDEs, activation of p38 mitogen-activated kinase (MAPK), fetal cell senescence, and inflammatory biomarkers (COX-2) produced in reproductive tissues. The results showed that both BDE-47 and BDE-99 induced ROS within 2 min and caused cell death-associated morphologic changes. BDE-99 activated p38 MAPK and increased the expression of cox-2. Therefore, without considering the etiologies, PBDEs may increase the risk of preterm birth by promoting both inflammatory response and oxidative stress [58].

5.9 Developmental Retardation

A nested case-control study of PBDE exposure and neonatal intrauterine growth retardation in Shanghai, China, showed that exposure to PBDEs during pregnancy can increase the risk of neonatal intrauterine growth retardation. Other studies have shown a negative correlation between PBDE exposure and neonatal birth weight, body length, and chest circumference. Lopez-Espinosa et al. determined the concentration of PBDEs in the serum of 670 Spanish pregnant women at gestational week 12 and in 534 umbilical cord samples, and the results showed that the PBDE levels were negatively correlated with abdominal circumference, estimated fetal weight, and biparietal diameter at weeks 20–34 and with birth, weight, and head circumference at delivery [59]. In addition, DNA methylation in human placenta may be related to intrauterine exposure to PBDEs, and changes in DNA methylation in human placenta (such as IGF2 and LINE1) may play an important role in pregnancy exposure to PBDEs and fetal growth [60].

Herbstman et al. tracked and analyzed the levels of PBDEs in cord blood of 329 pregnant women and the neurodevelopment of their children within 6 years after birth through a prospective cohort study. The results showed that children with higher concentrations of PBDEs scored lower on tests of mental and physical development at 12–48 and 72 months. Associations were significant for 12-month Psychomotor Development Index (BDE-47), 24-month Mental Development Index (MDI) (BDE-47, 99, and 100), 36-month MDI (BDE-100), 48-month full-scale and verbal IQ (BDE-47, 99, and 100) and performance IQ (BDE-100), and 72-month performance IQ (BDE-100), indicating neurodevelopmental effects in relation to cord blood PBDE concentrations [61].

5.10 Children Behavioral Problems

Behavioral problems in childhood may be a harbinger of adolescent mental problems, as well as poor academic performance, disciplinary violations, and physical and mental subhealth in adulthood. Evidence from existing epidemiological studies shows that FRs affect children's behavioral development. PBDE exposure during pregnancy is associated with executive dysfunction and poor sustained attention control in children. CHAMACOS Study and family studies reported that exposure to PBDEs in children may increase inattention. The results from a large prospective population showed that prenatal and postnatal exposure to PBDEs increased hyperactivity and aggression in children. In Menorca's cohort study, a higher risk of poor social competence symptoms was observed as a consequence of postnatal BDE-47 exposure [62].

Individuals with autistic disorder (ASD) experience varying degrees of difficulty in many social behaviors, such as communicating, helping each other, and maintaining relationships. They also exhibit repetitive and stereotypical behaviors, including repetitive movements, sensory behavior, and cognitive rigidity [63]. In the Health Outcomes and Measures of the Environment (HOME) Study, the Social Response Scale (SRS) score of children aged 4–5 years decreased with the increase of serum BDE-28 concentration in their mothers, indicating that increased levels of BDE-28 were associated with more autistic behaviors [64].

In addition, Hertz-Picciotto et al. [65] detected the blood samples of 100 children with autism/autism spectrum disorder, with developmental delay and from the general population, and found that children with autism/autism spectrum disorder and developmental delay were similar to typically developing controls for all PBDE congeners, but the levels were high for all three groups. However, plasma samples collected post-diagnosis in their study may not represent early life exposures due to changes in diet and introduction of new household products containing PBDEs. Studies with direct measurements of prenatal or infant exposures are needed to assess the possible causal role for these compounds in autism spectrum disorders.

References

1. Iqbal M, Syed JH, Katsoyiannis A et al (2017) Legacy and emerging flame retardants (FRs) in the freshwater ecosystem: a review [J]. Environ Res 152:26–42
2. Cristale J, Katsoyiannis A, Sweetman AJ et al (2013) Occurrence and risk assessment of organophosphorus and brominated flame retardants in the river Aire (UK) [J]. Environ Pollut 179:194–200
3. Garcia-Lopez M, Rodriguez I, Cela R (2010) Mixed-mode solid-phase extraction followed by liquid chromatography-tandem mass spectrometry for the determination of tri- and di-substituted organophosphorus species in water samples [J]. J Chromatogr A 1217(9):1476–1484
4. Martinez-Carballo E, Gonzalez-Barreiro C, Sitka A et al (2007) Determination of selected organophosphate esters in the aquatic environment of Austria [J]. Sci Total Environ 388(1–3):290–299
5. Bogdal C, Schmid P, Kohler M et al (2008) Sediment record and atmospheric deposition of brominated flame retardants and organochlorine compounds in Lake Thun, Switzerland: lessons from the past and evaluation of the present [J]. Environ Sci Technol 42(18):6817–6822
6. Cristale J, Garcia Vazquez A, Barata C et al (2013) Priority and emerging flame retardants in rivers: occurrence in water and sediment, daphnia magna toxicity and risk assessment [J]. Environ Int 59:232–243
7. Liu R, Nelson DO, Hurley S et al (2016) Association between serum polybrominated diphenyl ether levels and residential proximity to solid-waste facilities [J]. Environ Sci Technol 50(7):3945–3953
8. Luigi V, Giuseppe M, Claudio R (2015) Emerging and priority contaminants with endocrine active potentials in sediments and fish from the River Po (Italy) [J]. Environ Sci Pollut Res Int 22(18):14050–14066
9. Zhang X, Diamond ML, Robson M et al (2011) Sources, emissions, and fate of polybrominated diphenyl ethers and polychlorinated biphenyls indoors in Toronto, Canada [J]. Environ Sci Technol 45(8):3268–3274
10. Ni K, Lu Y, Wang T et al (2013) A review of human exposure to polybrominated diphenyl ethers (PBDEs) in China [J]. Int J Hyg Environ Health 216(6):607–623
11. Hites RA (2004) Polybrominated diphenyl ethers in the environment and in people: a meta-analysis of concentrations [J]. Environ Sci Technol 38(4):945–956
12. Jianxian S, Hui P, Jianying H (2015) Temporal trends of polychlorinated biphenyls, polybrominated diphenyl ethers, and perfluorinated compounds in Chinese sturgeon (Acipenser sinensis) eggs (1984-2008) [J]. Environ Sci Technol 49(3):1621–1630
13. Gabrielsen KM, Krokstad JS, Villanger GD et al (2015) Thyroid hormones and deiodinase activity in plasma and tissues in relation to high levels of organohalogen contaminants in East Greenland polar bears (Ursus maritimus) [J]. Environ Res 136:413–423
14. Persson S, Magnusson U (2015) Environmental pollutants and alterations in the reproductive system in wild male mink (Neovison vison) from Sweden [J]. Chemosphere 120:237–245
15. Zhao Y, Song Q, Cao Z et al (2018) Umbilical cord blood PBDEs concentrations in relation to placental size at birth [J]. Chemosphere 201:20–24
16. Zhang J, Chen L, Xiao L et al (2017) Polybrominated diphenyl ether concentrations in human breast milk specimens worldwide [J]. Epidemiology 28(Suppl 1):S89–s97
17. Chen Y, Wang X, Li Y et al (2015) Persistent organic pollutants in matched breast milk and infant faeces samples [J]. Chemosphere 118:309–314
18. Zhang X, Zhang K, Yang D et al (2014) Polybrominated biphenyl ethers in breast milk and infant formula from Shanghai, China: temporal trends, daily intake, and risk assessment [J]. Sci Total Environ 497-498:508–515
19. Shin MY, Lee S, Kim HJ et al (2016) Polybrominated diphenyl ethers in maternal serum, breast milk, umbilical cord serum, and house dust in a south korean birth panel of mother-neonate pairs [J]. Int J Environ Res Public Health 13(8):767

20. Guillette LJ Jr, Iguchi T (2012) Ecology. Life in a contaminated world [J]. Science 337(6102):1614–1615
21. Saillenfait AM, Ndaw S, Robert A et al (2018) Recent biomonitoring reports on phosphate ester flame retardants: a short review [J]. Arch Toxicol 92(9):2749–2778
22. Zhao F, Kang Q, Zhang X et al (2019) Urinary biomarkers for assessment of human exposure to monomeric aryl phosphate flame retardants [J]. Environ Int 124:259–264
23. Hoffman K, Fang M, Horman B et al (2014) Urinary tetrabromobenzoic acid (TBBA) as a biomarker of exposure to the flame retardant mixture Firemaster(R) 550 [J]. Environ Health Perspect 122(9):963–969
24. Bowers WJ, Wall PM, Nakai JS et al (2015) Behavioral and thyroid effects of in utero and lactational exposure of Sprague-Dawley rats to the polybrominated diphenyl ether mixture DE71 [J]. Neurotoxicology and teratology 52(Pt B):127–142
25. Cao J, Lin Y, Guo LH et al (2010) Structure-based investigation on the binding interaction of hydroxylated polybrominated diphenyl ethers with thyroxine transport proteins [J]. Toxicology 277(1–3):20–28
26. Lu D, Jin Y, Feng C et al (2017) Multi-analyte method development for analysis of brominated flame retardants (BFRs) and PBDE metabolites in human serum [J]. Anal Bioanal Chem 409(22):5307–5317
27. Ahmed OM, El-Gareib AW, El-Bakry AM et al (2008) Thyroid hormones states and brain development interactions [J]. Int J Dev Neurosci 26(2):147–209
28. Boas M, Feldt-Rasmussen U, Skakkebaek NE et al (2006) Environmental chemicals and thyroid function [J]. Eur J Endocrinol 154(5):599–611
29. Hamers T, Kamstra JH, Sonneveld E et al (2006) In vitro profiling of the endocrine-disrupting potency of brominated flame retardants [J]. Toxicol Sci 92(1):157–173
30. Gosavi RA, Knudsen GA, Birnbaum LS et al (2013) Mimicking of estradiol binding by flame retardants and their metabolites: a crystallographic analysis [J]. Environ Health Perspect 121(10):1194–1199
31. Butt CM, Stapleton HM (2013) Inhibition of thyroid hormone sulfotransferase activity by brominated flame retardants and halogenated phenolics [J]. Chem Res Toxicol 26(11):1692–1702
32. Patisaul HB, Roberts SC, Mabrey N et al (2013) Accumulation and endocrine disrupting effects of the flame retardant mixture Firemaster(R) 550 in rats: an exploratory assessment [J]. J Biochem Mol Toxicol 27(2):124–136
33. Hoffman K, Sosa JA, Stapleton HM (2017) Do flame retardant chemicals increase the risk for thyroid dysregulation and cancer? [J]. Curr Opin Oncol 29(1):7–13
34. Vazzana N, Santilli F, Sestili S et al (2011) Determinants of increased cardiovascular disease in obesity and metabolic syndrome [J]. Curr Med Chem 18(34):5267–5280
35. Prasad H, Ryan DA, Celzo MF et al (2012) Metabolic syndrome: definition and therapeutic implications [J]. Postgrad Med 124(1):21–30
36. Sherling DH, Perumareddi P, Hennekens CH (2017) Metabolic syndrome [J]. J Cardiovasc Pharmacol Ther 22(4):365–367
37. Darbre PD (2017) Endocrine disruptors and obesity[J]. Curr Obes Rep 6(1):18–27
38. Chen X, Huang C, Wang X et al (2012) BDE-47 disrupts axonal growth and motor behavior in developing zebrafish [J]. Aquat Toxicol 120-121:35–44
39. An J, Li S, Zhong Y et al (2011) The cytotoxic effects of synthetic 6-hydroxylated and 6-methoxylated polybrominated diphenyl ether 47 (BDE47) [J]. Environ Toxicol 26(6):591–599
40. Kuriyama SN, Talsness CE, Grote K et al (2005) Developmental exposure to low dose PBDE 99: effects on male fertility and neurobehavior in rat offspring [J]. Environ Health Perspect 113(2):149–154
41. Abdelouahab N, Ainmelk Y, Takser L (2011) Polybrominated diphenyl ethers and sperm quality [J]. Reprod Toxicol 31(4):546–550
42. Main KM, Kiviranta H, Virtanen HE et al (2007) Flame retardants in placenta and breast milk and cryptorchidism in newborn boys [J]. Environ Health Perspect 115(10):1519–1526

43. Chao HR, Shy CG, Wang SL et al (2010) Impact of non-occupational exposure to polybrominated diphenyl ethers on menstruation characteristics of reproductive-age females [J]. Environ Int 36(7):728–735
44. Li H, Su G, Zou M et al (2015) Effects of tris(1,3-dichloro-2-propyl) phosphate on growth, reproduction, and gene transcription of daphnia magna at environmentally relevant concentrations [J]. Environ Sci Technol 49(21):12975–12983
45. Ashwood P, Schauer J, Pessah IN et al (2009) Preliminary evidence of the in vitro effects of BDE-47 on innate immune responses in children with autism spectrum disorders [J]. J Neuroimmunol 208(1–2):130–135
46. Fernie KJ, Mayne G, Shutt JL et al (2005) Evidence of immunomodulation in nestling American kestrels (Falco sparverius) exposed to environmentally relevant PBDEs [J]. Environ Pollut 138(3):485–493
47. Saquib Q, Siddiqui MA, Ahmed J et al (2016) Hazards of low dose flame-retardants (BDE-47 and BDE-32): influence on transcriptome regulation and cell death in human liver cells [J]. J Hazard Mater 308:37–49
48. Araki A, Saito I, Kanazawa A et al (2014) Phosphorus flame retardants in indoor dust and their relation to asthma and allergies of inhabitants [J]. Indoor Air 24(1):3–15
49. Betts KS (2015) Tracking alternative flame retardants: hand-to-mouth exposures in adults [J]. Environ Health Perspect 123(2):A44
50. Darnerud PO (2003) Toxic effects of brominated flame retardants in man and in wildlife [J]. Environ Int 29(6):841–853
51. Hardell L, Carlberg M, Hardell K et al (2007) Decreased survival in pancreatic cancer patients with high concentrations of organochlorines in adipose tissue [J]. Biomed Pharmacother 61(10):659–664
52. Li ZH, Liu XY, Wang N et al (2012) Effects of decabrominated diphenyl ether (PBDE-209) in regulation of growth and apoptosis of breast, ovarian, and cervical cancer cells [J]. Environ Health Perspect 120(4):541–546
53. Hardell L, Lindstrom G, Van Bavel B et al (1998) Concentrations of the flame retardant 2,2′,4,4′-tetrabrominated diphenyl ether in human adipose tissue in swedish persons and the risk for non-Hodgkin's lymphoma[J]. Oncol Res 10(8):429–432
54. Schreder ED, Uding N, La Guardia MJ (2016) Inhalation a significant exposure route for chlorinated organophosphate flame retardants [J]. Chemosphere 150:499–504
55. Leonetti C, Butt CM, Hoffman K et al (2016) Brominated flame retardants in placental tissues: associations with infant sex and thyroid hormone endpoints [J]. Environ Health 15(1):113
56. Zhao Y, Ruan X, Li Y et al (2013) Polybrominated diphenyl ethers (PBDEs) in aborted human fetuses and placental transfer during the first trimester of pregnancy [J]. Environ Sci Technol 47(11):5939–5946
57. Peltier MR, Koo HC, Getahun D et al (2015) Does exposure to flame retardants increase the risk for preterm birth? [J]. J Reprod Immunol 107:20–25
58. Behnia F, Peltier MR, Saade GR et al (2015) Environmental pollutant polybrominated diphenyl ether, a flame retardant, induces primary amnion cell senescence [J]. Am J Reprod Immunol 74(5):398–406
59. Lopez-Espinosa MJ, Costa O, Vizcaino E et al (2015) Prenatal exposure to polybrominated flame retardants and fetal growth in the INMA cohort (Spain) [J]. Environ Sci Technol 49(16):10108–10116
60. Zhao Y, Liu P, Wang J et al (2016) Umbilical cord blood PBDEs concentrations are associated with placental DNA methylation [J]. Environ Int 97:1–6
61. Herbstman JB, Sjodin A, Kurzon M et al (2010) Prenatal exposure to PBDEs and neurodevelopment [J]. Environ Health Perspect 118(5):712–719
62. Gascon M, Vrijheid M, Martinez D et al (2011) Effects of pre and postnatal exposure to low levels of polybromodiphenyl ethers on neurodevelopment and thyroid hormone levels at 4 years of age [J]. Environ Int 37(3):605–611

63. Battle DE (2013) Diagnostic and statistical manual of mental disorders (DSM) [J]. CoDAS 25(2):191–192
64. Braun JM, Kalkbrenner AE, Just AC et al (2014) Gestational exposure to endocrine-disrupting chemicals and reciprocal social, repetitive, and stereotypic behaviors in 4- and 5-year-old children: the HOME study [J]. Environ Health Perspect 122(5):513–520
65. Hertz-Picciotto I, Bergman A, Fangstrom B et al (2011) Polybrominated diphenyl ethers in relation to autism and developmental delay: a case-control study [J]. Environ Health 10(1):1

Perfluorinated Compound Exposure and Health Effects in Humans

Jialin Li and Renshan Ge

Abstract Perfluorinated compounds (PFCs) include a large group of anthropogenic fluorinated organic substances. This chapter reviews the current understanding of the sources and trends of human exposure to PFCs and evidences from both animal experiments and epidemiological studies about their effects on immunological and metabolic functions, neurodevelopment, reproduction, and children development. Humans can be exposed to PFCs by ingesting contaminated drinking water and seafood, inhaling polluted indoor air, and other contaminated media, the way of exposure is wide and undetectable. The structure and physicochemical properties of PFCs can greatly affect their environmental behaviors and the safety of emerging perfluorobutane sulfonates (PFBS) and other substances cannot be ignored. Therefore, it is necessary to study health effects of various PFCs and their alternatives, and analyze the different environmental behaviors of isomers and homologues of PFCs.

Keywords Perfluorinated compounds · Exposure pathways · Health effect

1 Introduction

Perfluorinated compounds (PFCs) include a large group of anthropogenic fluorinated organic substances characterized by a fully fluorinated carbon backbone (containing only C-F and C-C bonds), typically 4–14 in length, and a charged functional moiety (primarily carboxylate, sulfonate, or phosphonate). In the 1950s, American

J. Li
School of Public Health, Fudan University, Shanghai, China

Key Laboratory of Public Health Safety, Ministry of Education, Fudan University, Shanghai, China

R. Ge (✉)
The Second Affiliated Hospital and Yuying Children's Hospital of Wenzhou Medical University, Wenzhou, China
e-mail: renshan_ge@wmu.edu.cn

Y. Zhang (ed.), *Emerging Chemicals and Human Health*,
https://doi.org/10.1007/978-981-32-9535-3_9

3 M Company first produced PFCs by the electrochemical fluorine method. Their unique physicochemical properties such as oil and water repellence; resistance to thermal, chemical, or biological degradation; and extremely low surface tension make them suitable materials for many industrial applications. As a matter of fact during the last half century, they have been commonly used in the formulation of stain repellents, coatings, textiles, paints, waxes, polishes, electronics, fire-fighting foams, surfactants, emulsifiers, additives, food packaging and other products, and some PFCs have been even used as plasma substitutes in the biomedical field.

Among PFCs, perfluorooctane sulfonate (PFOS) and perfluorooctane carboxylic acid (PFOA) with a carbon chain length of 8 carbon atoms are the most widely used and the final transformation product of various PFCs precursor substances in the environment. During 1970–2002, approximately 96,000 t of perfluorooctane sulfonyl fluoride (PFOSF, a precursor of PFOS) and 26,500 t of related by-products and wastes were produced worldwide, and 6800–45,250 t were eventually released directly or indirectly into the environment. PFCs are extremely persistent and are one of the most difficult organic pollutants to decompose. No significant degradation was observed in the hydrolysis of PFOS at various temperatures and pH. PFOS has excellent stability in both aerobic and anaerobic environments, and extensive studies using various microorganisms and conditions have shown that PFOS does not show any signs of degradation. The only case where PFOS decomposition occurs is incineration under high temperature conditions. The existence of bio-enrichment causes the concentration of fluorochemicals in living organisms to increase, which is one of the reasons for increasing the biological toxicity of organs. Physicochemical properties of PFCs (carbon chain length, carbon chain end group type and whether they contain branches), environmental conditions (ecosystem composition, water temperature), pollutant concentration, and other factors will affect the process of enrichment in the organism and the spread in the food chain. Differences in biological species and physiological and biochemical parameters (body length, weight, and gender) can also lead to different effects. PFOS can accumulate in organic organisms, and its accumulation in organisms is hundreds or thousands of times higher than known POPs such as dioxin and organochlorine pesticides. There is a lot of evidence that aquatic food chain organisms have a strong enrichment effect on PFOS. The concentration of fish on PFOS is 500–12,000 times. Because of their widespread use, environmental distribution, and the potential of some PFCs to bioaccumulate into higher trophic levels in the food chain and to be an important source of human exposure, they have gained worldwide attention from the scientific and regulatory communities.

PFCs also have reproductive toxicity, mutagenic toxicity, developmental toxicity, neurotoxicity, immunotoxicity, and other toxicities and are a class of environmental pollutants with systemic multiorgan toxicity. Toxicities observed in toxicological studies include inhibiting the immune system, affecting mitochondrial metabolism, causing liver cell damage and germ cell damage, reducing reproduction and fertility, affecting late fetal development, interfering with gene expression and enzyme activity, disrupting cell membrane structure, and changing thyroid function. Due to the persistent and bioaccumulative nature of certain PFCs, C8-C14 PFCs and their sodium and ammonium salts were included in the candidate list of regulatory substances in the EU, and PFOS has been added to the persistent organic pollutants (POPs) list at the Stockholm Convention on Persistent Organic Pollutants in 2009.

With the cessation of production of PFOS and PFOA by 3 M in 2000 and the voluntary cessation of production by eight major companies in 2006, PFOS and PFOA slowly faded out of the market. C11-C14 perfluorinated carboxylic acids (PFCAs), PFOA, and ammonium perfluorooctanoate (APFO) as vPvB (very persistent and very bioaccumulative) chemicals were included in the Candidate List of Substances of Very High Concern under the European Chemicals Regulations during 2012–2013. Different regulations on PFCs in different periods have led to changes in PFC production and time- or regional-dependent environmental distribution. In the past decade, due to these regulatory interventions, human exposure to PFOS and PFOA has been declining in Western countries and Japan, and understanding of the adverse effects on human health has been rapidly advancing. In order to make up for the gap in the market, several new chemicals have been developed as an alternative. At the same time, the environmental literature reports an increase of new PFCs with a shorter chain length which are difficult to detect using standard methods. New evidence from animal experiments suggests that some of these alternative PFCs may be equally hazardous. To sum up, this part aims to summarize the exposure pathways of PFCs and potential adverse health effects.

2 Exposure Levels and Pathways

Figure 1 [1] briefly summarizes the pathways in which humans are exposed to PFCs. Humans can be exposed to PFCs by ingesting contaminated drinking water and seafood, inhaling indoor air, and exposure to other contaminating media. Because of their widespread application and environmental persistence, they have resulted in global occurrence in the water, air, soil, sediment, indoor dust, wild life, and even the polar ice sheet. Direct exposures to industrial products can be quickly phased out following shifts in chemical production, while indirect exposures persist over long timescales, which resulted from PFCs accumulation in the ocean, soil, marine food chains, and aqueous film-forming foam (AFFF) contamination of groundwater. Since the concentration of pollutants in the blood has a good correlation with the concentration in the human body, and the process of collecting blood samples does not cause harm to the human body, the blood becomes a good indicator of human exposure. In addition, studies have examined the presence of PFCs in umbilical cord, breast milk, urine, hair, nails, and so on, suggesting a lifelong exposure to PFCs in humans, from embryonic until adult stage. Evidence is sufficient to show the widespread presence of PFCs both in the external environment and in the internal environment. There are also some studies exploring the relationship between parental and progeny for PFCs, and the results show that low levels of PFCs in cord sera and an increase in concentration through the first months of infant life. Although the concentrations in breast milk are low, this intake leads to a body burden at the age of 6 months similar to (PFOS) or higher than (PFOA) that found

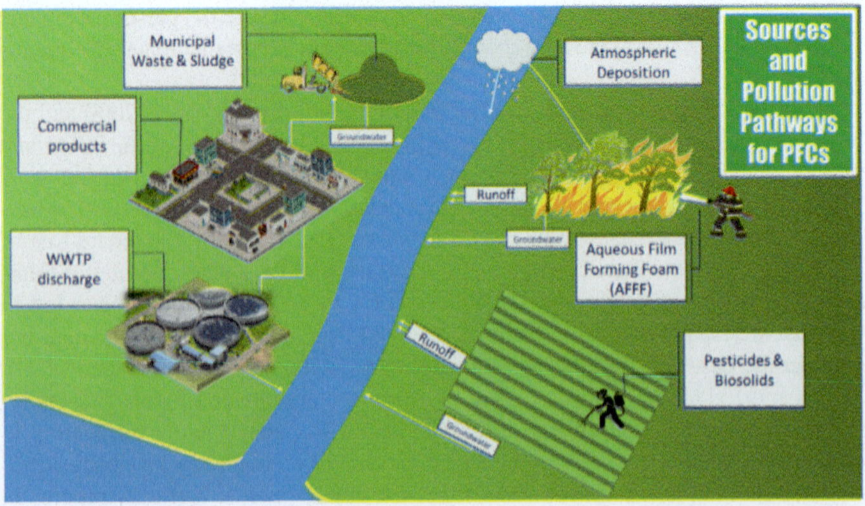

Fig. 1 Overview of the sources and pollution pathways of PFCs in the environment [1]

in adults. Understanding the relative importance of these different exposure pathways is thus particularly critical for analysis of the sources of PFCs in human biological samples, and for future exposure rating.

2.1 Air and Dust: PFCs Exposure

Due to low water solubility and high vapor pressure, neutral PFCs (mainly fluorotelomer alcohols (FTOHs), perfluorooctane sulfonamide (FOSA) and sulfonamidoethanols (FOSEs)) are the main components of PFCs in indoor air. FOSE compounds are the most commonly detected PFCs in indoor air, while the concentrations of FTOHs are much higher than the concentrations of other neutral PFCs. FTOHs are often used for the impregnation of furniture and flooring and are used as an intermediate in the manufacture of a variety of household consumer products such as paints, carpets, and cleaners. FOSA is an important PFOS precursor used for commercial applications. In general, the concentrations of FTOHs, FOSA, and FOSEs in indoor air are significantly lower in the office than at home. Unstable precursors can be detected in indoor environments when using products containing PFC precursors. These substances can be bio-converted into perfluoroalkyl acids (PFAAs) after inhalation in the body.

PFCs in indoor dust may be derived from high concentrations of FTOHs in indoor air, as FTHOs are their most important precursors. In general, changes in PFC concentrations in different countries may be attributed to different cultural traditions (such as the use of carpets or no carpets), architectural and decorative features, habits of purchasing clothes or consumer goods, sample quantities and sampling locations, sample collection, and analytical procedures and the economic status of a region. Although PFCs in various dusts are observed in offices, homes, and vehicles, indoor dust intake may not be an important route for human exposure compared to other pathways (food and drinking water).

With the elimination of PFOA and PFOS, the production and consumption of some new short-chain compounds have gradually increased, but toxicological studies on such alternative chemicals are still lacking. Sample analysis and detection play an important role in experimental research. Choosing an appropriate and effective method of analysis and detection brings great convenience to the research work. Common pretreatment methods for PFCs include solid phase extraction, liquid-liquid extraction, and ultrasonic extraction. With the rapid development and continuous improvement of mass spectrometry technology, especially the development of high-performance liquid chromatography tandem mass spectrometry technology, the main instrumental detection methods for PFCs at present are gas chromatography-mass spectrometry (GC-MS) and liquid chromatography-tandem mass spectrometry (LC-MS/MS). In order to detect more kinds of PFCs simultaneously and improve the detection accuracy of PFCs, researchers at home and abroad have made a lot of efforts in the optimization of sample pretreatment conditions, the selection of chromatographic columns, mobile phases and instrumental detection methods in recent years. However, Robel et al. [2] reported that only 16% of the PFCs detected by particle-induced gamma ray emission (PIGE) spectroscopy were detected using conventional methods. Therefore, the detection technology of modern chemicals should also be updated.

2.2 *Food: PFCs Exposure*

Some literature estimates were summarized in Table 1 [3] to characterize source contributions to overall PFC exposures for adults. Compared with inhalation or skin contact, dietary intake is widely believed to be the largest source of PFC exposure.

Fish and shellfish usually exhibit the highest PFC concentration and frequency of detection in all types of food. Some studies evaluated the relationship between PFCs concentration and human diet structure by collecting fish and shellfish consumption data around the world. The results showed that PFOS is the main PFCs in fish and shellfish, and other PFCs such as PFOA, perfluorononanoic acid (PFNA), perfluoroundecanoic acid (PFUnDA) and PFHxS also frequently appear in fish and shellfish at relatively high concentrations. The contribution of seafood to the PFC exposure of different populations is clearly different. Christensen et al. [4] found that fish and shellfish intakes were both associated with elevated levels of multiple PFCs with differences by types of fish or shellfish in the US National Health and Nutrition Exam Survey between 2007 and 2014. Hu et al. [5] showed that the presence of elevated serum concentrations of PFCs with $C \geq 9$ chain length in humans was useful for identifying when seafood was a dominant exposure source. Birth cohort data from the Faroe Islands [6] confirmed this observation by showing strong

Table 1 Estimates of source contributions (%) to adult PFC exposures [3]

PFCs	Diet	Dust	Food package	Tap water	Inhalation	Dermal	other	Ref.
PFOA	16	11	56		14		2[a]	[7]
	41			37			22[d]	[8]
	47	8		12	6		27[d]	[9]
	66	9		24	<1	<1		[10]
	77	8		11	4			[11]
	85	6	3[b]	1			4[c]	[12]
	99			<1				[13]
PFOS	66	10		7	2		16[d]	[14]
	72	6		22	<1	<1		[9]
	81	15					4[a]	[7]
	93			4			3[d]	[11]
	96	1		1	2			[12]
	100			<1				[13]
PFBA		4		96				[13]
PFDA	51	2		4	15		28[d]	[13]
PFHxA	38	4		38	8		12[d]	[13]
PFDoDA	86	2		2	4		5[d]	[13]

[a]Carpet
[b]Consumer goods
[c]Precursors
[d]Indirect

associations between serum concentrations of PFUnDA (C11) and hair mercury concentrations, a strong tracer of seafood consumption. Concentrations of traditional PFCs in marine biota have lagged behind the production changes of the compounds, leading to an increased significance of seafood as an exposure source.

Especially in Western countries, dairy products and beverages are also a source of PFCs because of the high consumption. However, only certain types of PFCs are detected in dairy products and beverages, and typically have low concentrations and low detection frequencies. The PFC content in milk is higher than that in yogurt, butter/cheese and beverages. Loading of different packaging materials in different regions may be a key factor affecting the concentration of PFCs in dairy products and beverages. The PFCs in beverages are mainly short-chain, such as PFBA and PPFEA. Short-chain PFCs are easier to detect in foods with higher moisture content than long-chain PFCs, probably because short-chain PFCs have higher solubility in foods with high moisture content, or they have higher soils. The higher plant absorption potential, the lower adsorption capacity in the root system. Elevated PFC concentrations in meat and dairy products have also been reported, suggesting that PFC uptake from biosolids contaminated agriculture is a source of dietary exposure for farm animals.

More concern about the significance of human exposures to PFCs originating from biosolids and agriculture is needed. The 2001 US EPA National Sewage Sludge Survey reported that, based on the 13 PFCs measured, the mean load of

∑PFCs in US biosolids was estimated at 2749–3450 kg/year, of which about 1375–2070 kg was applied on agricultural land and 467–587 kg went to landfills as an alternative disposal route. In many instances the leachate collected from landfills is sent back to the local wastewater and sludge from wastewater treatment plants (WWTP) and is thus re-incorporated into the sludge/liquid waste stream.

Wastewater is often recycled for agricultural irrigation, which is also a major source of PFCs that cannot be ignored. In one study [15], concentration factors of root to soil up to 4.7 and 10.3 were found for PFOS and PFOA, respectively, and all seven plants investigated displayed root concentration factors greater than 1, which certified the bioaccumulation of PFCs. Due to the discovery of PFCs in soil, irrigation water and air, vegetables may absorb PFCs from these media. Therefore, vegetables as an indispensable part of the human diet are also potential source of PFCs for humans. It has been reported that vegetables/fruits are an important source of food for PFOA intake (19%) in addition to drinking water. However, the PFCs detected in vegetables are different and the concentration is not high. The adsorption of organic matters by PFCs is positively correlated with the length of PFC chain, and the adsorption of sulfonate is stronger than that of carboxylate. Therefore, short-chain PFCs including carboxyl groups (PFHxA, PFHpA, and PFOA) are easily absorbed by plant roots, resulting in their widespread ubiquity in vegetables. Although PFOS is usually a main compound in the environment, its concentration in vegetables is generally lower than that of PFOA.

Perfluorochemicals are also used in the manufacturing of food-contact substances that represent potential sources of oral exposure to these chemicals. The most recognizable products are perfluorochemicals in nonstick coatings (polytetrafluoroethylene, PTFE) for cookware and in paper coatings for oil and moisture resistance. Analysis of PTFE cookware showed residual amounts of PFOA in the low μg/kg range. PFOA was present in microwave popcorn bag paper at amounts as high as 300 μg/kg [16]. During actual package use, PFCs do migrate from fluorochemical-treated paper into food simulants such as butter, water, vinegar, and water/ethanol mixtures.

2.3 *Drinking Water: PFCs Exposure*

Various PFCs are detected in different drinking water samples collected from various countries. PFOS, PFHxA, PFHpA, PFNA, PFDA, and PFOA appear to be the main PFCs with the highest concentrations and frequencies of detection in drinking water. The concentration of PFOA contamination in drinking water is generally 0–20 ng/L. The concentrations of PFCs rank from highest to lowest as well water > tap water > bottled water > drinking water > raw water. PFCs pollution in well water is the point source. PFCs concentrations in tap water and drinking water are higher than the concentration in raw water, suggesting that drinking water treatment processes can introduce new PFCs contamination. At the same time, water bottles and label inks may be a potential source of PFHpA for bottled water, but more direct evidence is needed to support this hypothesis.

Aquatic system is considered as the most important way for PFCs when they transfer into the surroundings. Generally, PFCs are transported into the aqueous environment via discharge from WWTP or fluorochemical industries, domestic consumption and atmospheric deposition. Table 2 shows the mean concentrations (pg/L) of PFCs in different types of aqueous environments. It is reported that drinking water accounts for 75% of total PFC emissions near contaminated sites. Although western manufacturers began a voluntary phase out of PFOS and PFOA production in 2002, both PFOS and PFOA persist in the environment because of extremely stable carbon-fluorine bonds and circulate through environmental media including air, soil, plants and food stocks. Meanwhile, as an increase of new short-chain PFC production, not only PFOS, PFOA, and precursors but also other modern PFCs may now accumulate in drinking water and are associated with human exposure. It is worth mentioning that in the samples of some nondeveloped countries in Africa and Asia, the concentration of PFCs is much higher than that of developed countries in Europe and America, and the socio-economic level is also related to the pollution level of PFCs. Although drinking water pollution is prevalent throughout the world, there is increasing research evidence showing population exposure and multiple health damages, but unfortunately there is no international/national PFAS drinking water standard.

3 Health Effects of PFCs in Humans

Due to the unique physical and chemical properties of PFCs, unlike the "classic" mode that other POPs accumulate in fat, PFCs preferentially bind to proteins in organisms and are most abundant in organs such as blood, liver, and muscle. According to the WHO classification criteria for teratogenic toxicity of compounds, PFOS and F-53B are moderately toxic compounds and belong to the third class of chemicals (harmful to aquatic organisms) [36]. At present, the development of aquatic ecotoxicology of PFCs is relatively mature. Although the doses of PFOS and PFOA used in most of the studies on toxic effects of experimental animals far exceed the actual exposure level of the general population, studies have shown that PFC exposure at actual environmental concentration also affects human health. However, PFCs are still at a relatively early stage in the study of population health risks. Pressure to phase out some PFCs, such as PFOS and PFOA, has led to precipitous increases in the production of unstudied and unregulated novel replacement compounds such as perfluoroether carboxylic acids (e.g., GenX, Adona), chlorinated polyfluoroether sulfonates (e.g., F-53B), and fluorotelomer alcohols (e.g., Novec 1230). Since widespread exposure is reported, the impact of PFCs on human health is recently gaining awareness.

Table 2 Mean concentrations (pg/L) of PFCs in different types of aqueous environments

Occurrence	Region	Sampling period	TFA	PFBA	PFPA	PFHxA	PFHpA	PFOA	PFNA	PFDA	PFBS	PFHxS	PFOS
Sea/Ocean	Coastal area of Korea [17]	2002–2004						5700					1300
	Mediterranean Sea [18]	2005	500–50,000										
	Shandong peninsula, China [19]	2012		2100	2900	9500	2900	35,600	2600	1900	1100	3400	12,900
Rivers	River Elbe, Germany [20]	2007			2000	3300	1600	6200	700	400	2600	1100	1700
	Hanjiang River, China [21]	2010						41,000	9500	13,200			26,100
	Lambro River, Italy [22]	2011–2012		26,000		31,000	47,500	159,000	86,000	52,000	33,000	2500	27,000
	Ganges River, India [23]	2014			1475	2171	1229	403	76	68	5554	120	507
	River water, the Netherlands [24]	2016		5400	4500	6200	1900	4700	710	410	22,000	2100	3600
Rain and snow water	Canada [25]	2002	110,000										
	Tsukuba City and Kawaguchi City, Japan [26]	2007	65,000	900		700	1000	1100	2000	550			500

(continued)

Table 2 (continued)

Occurrence	Region	Sampling period	TFA	PFBA	PFPA	PFHxA	PFHpA	PFOA	PFNA	PFDA	PFBS	PFHxS	PFOS
Surface/ urban runoffs	Tokyo, Japan [27]	2005						90,000	43,000	22,000			50,000
	New York City, USA [28]	2006					900	5100		1050			1700
	Guangzhou, China [29]	2007–2008	225,000										
	Shanghai, China [30]	2015		5460		1734	3960	5627	1980	750	3861	1399	7190
Drinking water	Belpre, USA [31]	2002						95,000					
	Arnsberg-Neheim, Germany [31]	2006						600,000					
	Ruhr area, Germany [32]	2006				21,000		110,000			12,000		9000
	Blacktown, Australia [33]	2010				770	730	1640			740	3470	3920
	Drinking water, the Netherlands [24]	2016				3100		4000			6000		
Ground water	Military bases, USA [34]	2001–2014									60,000,000	250,000,000	100,000,000
	Ganges River, India [23]	2014			1475	2171	1229	403	76	68	5554	120	507
	AFFF-impacted site, USA [35]	2014–2016		3,700,000		7,600,000	1,500,000	10,700,000			1,800,000	21,000,000	33,100,000

3.1 Immunotoxicity

A study [37] on the effects of PFOA and PFOS on human peripheral leukocytes showed that PFOS induced a significant decrease in the activity of natural killer (NK) cells and reduced the release of phospholipid-stimulated anti-inflammatory cytokine TNF-α; PFOA significantly increased the differentiation of calcitriol-induced monocytes (HL-60 cell lineage). A corresponding animal study showed that repeated oral PFOS exposure reduced lymphocyte proliferation and NK cell activity in C57BL/6 mice, indicating an immunosuppressive effect of PFOS.

Besides, immunotoxicity of PFCs has been demonstrated in multiple animal models, including rodents, birds, reptiles, and other mammalian and nonmammalian wildlife. Sunderland et al. [3] showed findings from 25 epidemiological studies published between 2008 and 2018. Among five studies that examined the association between PFC exposure and suppression of antibody response to vaccination among children, adolescents or adults, four studies found statistically significant associations. The findings from a Faroese birth cohort [38] supported the notion that the developing adaptive immune system was particularly vulnerable to immunotoxicity during infancy. This vulnerability appeared to be the greatest during the first 6 months after birth, where PFC exposures were affected by breast-feeding.

By using primary human decidual stromal cells of early pregnancy, Yang et al. [39] demonstrated that PFOS inhibited decidualization of the stromal cells as well as decidualization-induced upregulation of 11β-hydroxysteroid dehydrogenase 1 (11β-HSD1), an enzyme that regenerates biologically active cortisol from its inactive counterpart cortisone. Moreover, PFOS attenuated cortisol-induced decidualization and upregulation of 11β-HSD1 in the stromal cells. Furthermore, PFOS inhibited the reduction of IL-6 and IL-1β, the key proinflammatory cytokines in maternal-fetal immune intolerance, by cortisone in the decidual stromal cells, indicating attenuated conversion of cortisone to cortisol. In conclusion, exposure to PFOS may disrupt the regeneration of cortisol in the decidual tissue, thereby impairing the decidualization and immune-tolerance environment of early pregnancy.

Except molecular-level (i.e., antibody concentrations), organ/system-level (i.e., infection of the respiratory system) also can reflect the health outcomes related to PFAS immunotoxicity. Since infection and allergy can be caused by food and airborne allergens, it is challenging to identify the contribution of PFC exposures in a low signal-to-noise setting. Up to now, results with organ/system-level outcomes such as asthma, infection, and allergies are more inconsistent.

3.2 Metabolic Effects

Biomarkers associated with PFC exposure in serum participate in a variety of important metabolic pathways associated with oxidative/nitrosative stress and exposure to low levels of environmental PFCs may increase the risk of various diseases, including metabolic syndrome, diabetes, and cardiovascular disease.

In a comprehensive review, Steenland et al. [40] evaluated the epidemiologic literature on PFOA and its association with several health outcomes. They finally drew a conclusion that the most consistent health findings were for modest increases in cholesterol may increase the risk of cardiovascular diseases, and also increase the level of serum uric acid. The mechanism by which PFCs are related to changes in serum cholesterol in humans is unknown, but the finding of positive associations between PFCs and altered cholesterol levels contradicts what would be expected from animal studies. In animals, PFOA was a strong peroxisome proliferator activated receptor alpha (PPARα) activator in the liver and this proliferation altered lipid metabolism with decreases, not increases, in serum lipids [41].In humans, the possible interference of PFCs with metabolism of fatty acids should not be underestimated and deserves further investigations on the mechanisms of action and as a possible risk factor for metabolic disorders and/or cardiovascular diseases. Certain chronic diseases with an inflammatory component, including diabetes, heart disease, and stroke, have been reported to be elevated in studies of exposed populations; however, the evidence is too meager and inconsistent for solid conclusions (Table 3).

3.3 *Neurodevelopmental Effects*

A cross-sectional study conducted by Gump et al. in the United States showed significant correlations between the blood levels of PFCs and severity of response inhibition impairments in children aged 9–11 years. In addition, the exposure levels of PFCs and impulsive behavioral characteristics were highly correlated [42].

A prospective cohort study on the levels of PFCs in the maternal cord blood and neurodevelopmental development in children conducted by Chen et al. [43] showed that the level of PFOS in the maternal cord blood was significantly negatively associated with the behavioral responsiveness of children at 2 years old. In the assessment of children's various types of behavior using a comprehensive scoring method, PFC exposure was inversely correlated with the establishment level of children's social and self-reliance skills [44].

Meanwhile, a prospective cohort study on maternal serum PFOS and PFOA levels and neurodevelopmental performance in infants at 6 months and 18 months of age was initiated by Goudarzi et al. [45] in Japan. In this research, a measure of specific response to neuromotor activity and development was used. The results showed that there was a significant correlation between the level of PFOA in maternal serum and the mental development index of the female babies at 6 months of age, indicating that maternal PFC exposure may have a certain influence on the neurodevelopment of infants.

Hoffman et al. [46] conducted a cross-sectional study in the United States to measure the PFC levels in sera of 571 children aged 12–15 years, and 48 of these children had been diagnosed with attention deficit hyperactivity disorder (ADHD). After logistic regression analysis and Spearman correlation analyses, the results

Table 3 Epidemiologic studies on PFAS exposures and metabolic outcomes [3]

Outcome	# of total studies	# of studies by results				Other PFCs
		PFOA	PFNA	PFHxS	PFOS	
Lipid profile[a]	39	21/10/1[b]	8/1/2	4/4/2	20/9/3	Inconsistent results for PFDA, PFUnDA, PFTeDA
Insulin resistance and diabetes	18	6/9/1	3/5/0	1/2/1	7/4/1	Mostly null for PFDA, PFUnDA, PFDoDA, N-EtFOSAA, N- MeFOSAA; one positive finding for PFDoDA and insulin resistance
Hypertension, vascular disease and stroke	10	3/5/1	3/0/1	0/3/1	1/3/1	Only one study reported null for PFDA and PFUnDA
Cardiovascular disease	6	1/4/1	1/0/0	0/10	0/1/0	No other PFASs have been investigated
Uric acid	5	4/0/0	0/0/0	0/1/0	2/2/0	No other PFASs have been investigated
Overweight and obese	4	1/3/0	1/1/0	1/1/0	3/1/0	Positive finding for PFDA in only one study

[a]Lipid profile includes low-density lipoprotein cholesterol (LDL-C), high-density lipoprotein cholesterol (HDL-C), total cholesterol, and triglycerides
[b]Number of studies with adverse/null/protective results

showed that the exposure levels of PFCs in sera of these children were significantly positively correlated with the probability of children with ADHD, indicating that, to some extent, there is a certain correlation between PFC exposure and ADHD disease in view of environmental factors.

The above studies indicate that the level of PFCs in vivo does have a certain correlation with the neurodevelopmental status of children and adolescents, and there is an exposure-effect relationship between PFCs and ADHD. That is to say ADHD is likely to be a typical healthy outcome of a neurotoxic disease caused by high exposure to PFCs.

3.4 Reproductive and Developmental Toxicity

PFCs have long been considered to be associated with adverse reproductive health effects. Studies have shown that exposure to perfluorochemicals is negatively correlated with sperm quality in men. Animal experiments showed that PFOA at a certain concentration could cause oxidative damage to testis and affect the activity of sorbitol dehydrogenase (SDH) and lactate dehydrogenase (LDH), which are the markers of energy supply during spermatogenesis, thus affecting the quantity, quality, and activity of sperm in mice. PFOS and PFOA were shown to reduce testosterone production in mice or rats and PFOS was associated with reduced epididymal sperm counts in mice. As far as human studies are concerned, a study among human workers at a PFOA-producing facility found no association between PFOA exposure and reproductive hormones [47].

The INUENDO cohort [48] is the largest one to date investigating the associations between PFC exposure and human semen quality. The most robust finding in this study was the negative associations between PFOS exposure and sperm morphology suggesting adverse effects of PFOS on semen quality, possibly due to interference with the endocrine activity or sperm membrane function. Two previous studies have been published evaluating the semen quality in relation to PFC exposure among 105 young Danish men [49] and 256 American infertility clients [50]. In these studies no statistically significant association between PFC exposures and sperm concentration, volume, or motility were detected, but in the Danish study, a combined estimate of high PFC exposure based on the sum of quartile scores of PFOS and PFOA indicated that high PFC exposure was associated with a lower proportion of normal sperm [49]. Sperm cell morphology was unfortunately not evaluated in the American study [50]. The finding [48] of a significant association between PFOS exposure and the proportion of normal sperm is in line with results from the previous Danish study [49] and further indicates that the association between PFC exposure and normal sperm is more likely to be caused by exposure to PFOS than to PFOA.

In addition to males, toxic effects have also been found in females. Animal experiments [51] found that PFOA had a lasting effect on the mammary glands of female offspring, and early exposure to PFOA after birth may permanently alter mammary gland development. PFOA also delayed puberty time in female mice. PFDoA did not necessarily affect puberty in rats, but had an effect on steroidogenic enzymes as well as E2 production and E2 receptors. Epidemiological studies showed that exposure to PFAS seriously damaged human reproductive health, whether in fertility or infertility. Women with higher serum PFOA levels had higher parity and longer pregnancy. Women with high exposure to perfluorooctanoate had an abnormal menstrual cycle rate. Even low levels of exposure to PFCs reduced fertility, and environmental exposure to PFCs impaired a woman's reproductive capacity by delaying pregnancy time.

In rodents, gestational exposure to PFOA, PFOS, or PFNA was found to be associated with fetal mortality, birth weight reduction, and delays in maturation. However, the results of human epidemiologic studies are equivocal.

In the Taiwan Birth Panel Study (TBPS) [43], only PFOS levels were found to be inversely associated with gestational age, birth weight, and head circumference [per ln unit: adjusted β (95% confidence interval, CI) = −0.37 (−0.60, −0.13) wks, −110.2 (−176.0, −44.5) g and − 0.25 (−0.46, −0.05) cm]. Additionally, the odds ratio of preterm birth, low birth weight, and small for gestational age increased with PFOS exposure level [per ln unit: adjusted odds ratio (OR) (95%CI) = 2.45 (1.47, 4.08), 2.61 (0.85, 8.03) and 2.27 (1.25, 4.15)]. When PFOS levels were divided into quartiles, a dose-response relation was observed. However, PFOA, PFNA, and PFUA were not observed to have any convincing impact on birth outcomes.

Studies that investigated the impact of prenatal PFOS and PFOA exposure on birth weight reveal either negative associations or null effects. The results are similar for birth outcomes, such as birth length, head circumference and Ponderal index. However, their influences on gestational age, preterm birth, low birth weight and small for gestational age are more diverse.

3.5 Other Health Effects

The mean number of proliferating thyroid follicular cells was increased 2.1-fold over control in GD 20 female fetuses from 1.0 mg/kg-d-treated dams [52]. Besides thyroid, evidence also shows that PFCs affected organs like liver and kidney. Animal studies revealed that PFOS tended to accumulate in livers and was able to induce hepatomegaly. Since PFOS-elicited pathological manifestation resembles one of the most common human liver diseases, nonalcoholic fatty liver disease, environmental exposure to PFOS may attribute to the disease progression.

Stanifer et al. [53] identified 74 studies, including 21 epidemiologic, 13 pharmacokinetic, and 40 toxicological studies in relation to kidney health. Three population-based epidemiologic studies demonstrated associations between PFC exposure and lower kidney function. Along with toxicology studies (n = 10) showing tubular histologic and cellular changes from PFC exposure, pharmacokinetic studies (n = 5) demonstrated the kidneys were major routes of elimination, with active proximal tubule transport. In several studies (n = 17), PFC exposure altered several pathways linked to kidney disease, including oxidative stress pathways, peroxisome proliferators-activated receptor pathways, NF-E2-related factor 2 pathways, partial epithelial mesenchymal transition, and enhanced endothelial permeability through actin filament modeling.

4 Safety of Modern PFCs

After discontinuing the use of PFOS and PFOA substances, 3 M Company developed a new substitute represented by PFBS and evaluated the safety of this substance [54]. Compared with PFOS and PFOA, PFBS was administered orally. The experimental mice did not show reproductive and developmental toxicity. At the same time, 3 M also found that the half-life of PFBS in mice was much less than that of PFOS and PFOA.

Similarly, others have analyzed the safety of emerging PFBS and other substances. For example, Newsted et al. [55] compared the acute toxicity of PFBS and PFOS to wild ducks and North American cockroaches. It was found that PFBS did not exhibit acute toxicity like PFOS for birds, and for body weight and visceral weight, the change of PFBS has no big impact.

However, Slotkin et al. [56] found that PFBS also showed toxicity to undifferentiated nerve cells, although it did not affect DNA synthesis like PFOS and PFOA. However, it promoted the occurrence of lipid peroxidation, causing damage to nerve cells. Therefore, PFBS may have neurotoxicity.

Oldham et al. further studied the lipid peroxidation exhibited by PFBS [57]. The study results showed that PFBS affected the transition temperature of the bilayer phospholipid membrane, that is, it affected the transition of the cell membrane from the liquid crystal state to the gel state. The liquid crystal state is the normal state under the physiological condition of the cell membrane, so PFBS can affect the fluidity of the cell membrane, and finally show the membrane toxicity. At the same time, the article also pointed out that PFBS was toxic to phospholipids, the main component of pulmonary surfactants.

Moreover, another alternative, PFHxA, shows toxicity in animal experiments [58], such as weight loss in mice, red blood cell count, ALT, AST, and ALP, and liver weight increase in male mice. Moreover, PFHxA is more toxic to aquatic animals than PFOA [59]. PFHxA is more rapidly metabolized in experimental animals than PFBS, and is very toxic at a low accumulation level.

Therefore, it can be seen that although new fluoride substitutes reduce bioaccumulation and toxicity, they may be difficult to degrade, and the characteristics of diffusion with water flow may cause a wider range of persistent pollution. Moreover, these new chemicals have not been fully verified for safety and can possibly cause damage to the human body. All these suggest that we need to be alert to such chemicals.

PFCs have very stable chemical properties. So far, no evidence and relevant reports have been found on their degradation in vivo, and no natural degradation pathway has been found. Conventional advanced oxidation technology cannot effectively degrade PFCs. At present, the main research methods are separation, removal, and chemical degradation. Separation and removal technologies of PFCs represented by PFOA and PFOS mainly include reverse osmosis and adsorption. Activated carbon and ion exchange resin are commonly used adsorbents for removing PFCs. In recent years, with the increasing attention paid to PFCs in the environment, researchers in various countries have developed a variety of new adsorbents for PFCs, such as chitosan and carbon nanotubes. The chemical degradation of PFOS and PFOA mainly includes thermal decomposition, oxidation degradation, reduction degradation, and direct photolysis. As a new type of POPs, PFCs have been studied by researchers at home and abroad, but still many problems have to be solved urgently. In previous studies where PFOS and PFOA are usually studied, other homologues are seldom studied. The structure and physicochemical properties of PFCs can greatly affect their environmental behaviors. Therefore, it is necessary to study various PFCs and analyze the different environmental behaviors of isomers and homologues of PFCs. At the same time, PFC removal technology, separation and chemical degradation, is highly costly and secondary pollution is easy to occur. Therefore, the research on PFC biodegradation technology (such as microorganisms and plants) should be strengthened.

5 Summary

This chapter reviews the general characteristics of PFCs, population exposure, human health and toxicological effects. According to the above summary, it is necessary to control and eliminate the exposure and adverse effects of PFCs. Solutions are needed to reduce the health and environmental hazards of PFCs, including reducing the use of perfluorochemical products (according to EU's new chemical regulation measures, starting from January 1, 2015, that the contents of ammonium pentafluorooctanoate or PFOA in products exceed concentration restrictions are banned for sale), actively looking for alternatives, exploring safe degradation pathways, strengthening biological monitoring, and striving to develop better adsorption devices in the future research.

References

1. Ateia M, Maroli A, Tharayil N et al (2019) The overlooked short- and ultrashort-chain poly- and perfluorinated substances: a review [J]. Chemosphere 220:866–882
2. Robel AE, Marshall K, Dickinson M et al (2017) Closing the mass balance on fluorine on papers and textiles [J]. Environ Sci Technol 51(16):9022–9032
3. Sunderland EM, Hu XDC, Dassuncao C et al (2019) A review of the pathways of human exposure to poly- and perfluoroalkyl substances (pfass) and present understanding of health effects [J]. J Exposure Sci Environ Epidemiol 29(2):131–147

4. Christensen KY, Raymond M, Blackowicz M et al (2017) Perfluoroalkyl substances and fish consumption [J]. Environ Res 154:145–151
5. Hu XDC, Dassuncao C, Zhang XM et al (2018) Can profiles of poly- and perfluoroalkyl substances (pfass) in human serum provide information on major exposure sources? [J]. Environ Health 17
6. Dassuncao C, Hu XC, Nielsen F et al (2018) Shifting global exposures to poly- and perfluoroalkyl substances (pfass) evident in longitudinal birth cohorts from a seafood-consuming population [J]. Environ Sci Technol 52:3738–3747
7. Trudel D, Horowitz L, Wormuth M et al (2008) Estimating consumer exposure to pfos and pfoa [J]. Risk Anal 28(2):251–269
8. Vestergren R, Cousins IT (2009) Tracking the pathways of human exposure to perfluorocarboxylates [J]. Environ Sci Technol 43(15):5565–5575
9. Haug LS, Huber S, Becher G et al (2011) Characterisation of human exposure pathways to perfluorinated compounds - comparing exposure estimates with biomarkers of exposure [J]. Environ Int 37(4):687–693
10. Lorber M, Egeghy PP (2011) Simple intake and pharmacokinetic modeling to characterize exposure of americans to perfluoroctanoic acid, pfoa [J]. Environ Sci Technol 45(19):8006–8014
11. Tian Z, Kim SK, Shoeib M et al (2016) Human exposure to per- and polyfluoroalkyl substances (pfass) via house dust in Korea: implication to exposure pathway [J]. Sci Total Environ 553:266–275
12. Shan GQ, Wang Z, Zhou LQ et al (2016) Impacts of daily intakes on the isomeric profiles of perfluoroalkyl substances (pfass) in human serum [J]. Environ Int 89-90:62–70
13. Gebbink WA, Berger U, Cousins IT (2015) Estimating human exposure to pfos isomers and pfca homologues: the relative importance of direct and indirect (precursor) exposure [J]. Environ Int 74:160–169
14. Egeghy PP, Lorber M (2011) An assessment of the exposure of americans to perfluorooctane sulfonate: a comparison of estimated intake with values inferred from nhanes data [J]. J Expo Sci Environ Epidemiol 21(2):150–168
15. Wen B, Wu Y, Zhang H et al (2016) The roles of protein and lipid in the accumulation and distribution of perfluorooctane sulfonate (pfos) and perfluorooctanoate (pfoa) in plants grown in biosolids-amended soils [J]. Environ Pollut (Barking, Essex: 1987) 216:682–688
16. Begley TH, Hsu W, Noonan G et al (2008) Migration of fluorochemical paper additives from food-contact paper into foods and food simulants [J]. Food Addit Contam 25(3):384–390
17. Yamashita N, Kannan K, Taniyasu S et al (2005) A global survey of perfluorinated acids in oceans [J]. Mar Pollut Bull 51(8–12):658–668
18. Scott BF, Macdonald RW, Kannan K et al (2005) Trifluoroacetate profiles in the arctic, Atlantic, and pacific oceans [J]. Environ Sci Technol 39(17):6555–6560
19. Wan Y, Wang SL, Cao XZ et al (2017) Perfluoroalkyl acids (pfaas) in water and sediment from the coastal regions of Shandong peninsula, China[J]. Environ Monit Assess 189(3):100
20. Ahrens L, Felizeter S, Sturm R et al (2009) Polyfluorinated compounds in waste water treatment plant effluents and surface waters along the river elbe, Germany [J]. Mar Pollut Bull 58(9):1326–1333
21. Wang BB, Cao MH, Zhu HD et al (2013) Distribution of perfluorinated compounds in surface water from hanjiang river in Wuhan, China [J]. Chemosphere 93(3):468–473
22. Castiglioni S, Valsecchi S, Polesello S et al (2015) Sources and fate of perfluorinated compounds in the aqueous environment and in drinking water of a highly urbanized and industrialized area in Italy [J]. J Hazard Mater 282:51–60
23. Sharma BM, Bharat GK, Tayal S et al (2016) Perfluoroalkyl substances (pfas) in river and ground/drinking water of the ganges river basin: emissions and implications for human exposure [J]. Environ Pollut 208:704–713
24. Gebbink WA, Van Asseldonk L, Van Leeuwen SPJ (2017) Presence of emerging per- and polyfluoroalkyl substances (pfass) in river and drinking water near a fluorochemical production plant in the Netherlands [J]. Environ Sci Technol 51(19):11057–11065

25. Scott BF, Spencer C, Mabury SA et al (2006) Poly and perfluorinated carboxylates in north american precipitation [J]. Environ Sci Technol 40(23):7167–7174
26. Taniyasu S, Kannan K, Yeung LWY et al (2008) Analysis of trifluoroacetic acid and other short-chain perfluorinated acids (c2-c4) in precipitation by liquid chromatography-tandem mass spectrometry: comparison to patterns of long-chain perfluorinated acids (c5-c18) [J]. Anal Chim Acta 619(2):221–230
27. Murakami M, Shinohara H, Takada H (2009) Evaluation of wastewater and street runoff as sources of perfluorinated surfactants (pfss) [J]. Chemosphere 74(4):487–493
28. Kim SK, Kannan K (2007) Perfluorinated acids in air, rain, snow, surface runoff, and lakes: relative importance of pathways to contamination of urban lakes [J]. Environ Sci Technol 41(24):8328–8334
29. Wang QY, Wang XM, Ding X (2014) Rainwater trifluoroacetic acid (tfa) in Guangzhou, South China: levels, wet deposition fluxes and source implication [J]. Sci Total Environ 468:272–279
30. Sun R, Wu MH, Tang L et al (2018) Perfluorinated compounds in surface waters of shanghai, China: source analysis and risk assessment [J]. Ecotoxicol Environ Saf 149:88–95
31. Paustenbach DJ, Panko JM, Scott PK et al (2007) A methodology for estimating human exposure to perfluorooctanoic acid (pfoa): a retrospective exposure assessment of a community (1951-2003) [J]. J Toxicol Environ Health-Part A-Curr Issues 70(1):28–57
32. Skutlarek D, Exner M, Farber H (2006) Perfluorinated surfactants in surface and drinking waters [J]. Environ Sci Pollut Res Int 13(5):299–307
33. Thompson J, Eaglesham G, Mueller J (2011) Concentrations of pfos, pfoa and other perfluorinated alkyl acids in australian drinking water [J]. Chemosphere 83(10):1320–1325
34. Barzen-Hanson KA, Field JA (2015) Discovery and implications of c-2 and c-3 perfluoroalkyl sulfonates in aqueous film-forming foams and groundwater [J]. Environ Sci Technol Lett 2(4):95–99
35. Xiao X, Ulrich BA, Chen B et al (2017) Sorption of poly- and perfluoroalkyl substances (pfass) relevant to aqueous film-forming foam (afff)-impacted groundwater by biochars and activated carbon [J]. Environ Sci Technol 51(11):6342–6351
36. Gsh (2015) Globally harmonized system of classification and labelling of chemicals(ghs).First revised edition. United Nations: Nations Publication:ST/SG/AC.10/30/Rev.6
37. Brieger A, Bienefeld N, Hasan R et al (2011) Impact of perfluorooctanesulfonate and perfluorooctanoic acid on human peripheral leukocytes [J]. Toxicol In Vitro 25(4):960–968
38. Grandjean P, Andersen EW, Budtz-Jorgensen E et al (2012) Serum vaccine antibody concentrations in children exposed to perfluorinated compounds[J]. JAMA 307(4):391–397
39. Yang Q, Wang W, Liu C et al (2016) Effect of pfos on glucocorticoid-induced changes in human decidual stromal cells in the first trimester of pregnancy [J]. Reprod Toxicol 63:142–150
40. Steenland K, Fletcher T, Savitz DA (2010) Epidemiologic evidence on the health effects of perfluorooctanoic acid (pfoa) [J]. Environ Health Perspect 118(8):1100–1108
41. Kennedy GL, Butenhoff JL, Olsen GW et al (2004) The toxicology of perfluorooctanoate[J]. Crit Rev Toxicol 34(4):351–384
42. Gump BB, Wu Q, Dumas AK et al (2011) Perfluorochemical (pfc) exposure in children: associations with impaired response inhibition [J]. Environ Sci Technol 45(19):8151–8159
43. Chen MH, Ha EH, Wen TW et al (2012) Perfluorinated compounds in umbilical cord blood and adverse birth outcomes [J]. PLoS One 7(8)
44. Chen MH, Ha EH, Liao HF et al (2013) Perfluorinated compound levels in cord blood and neurodevelopment at 2 years of age[J]. Epidemiology 24(6):800–808
45. Goudarzi H, Nakajima S, Ikeno T et al (2016) Prenatal exposure to perfluorinated chemicals and neurodevelopment in early infancy: the Hokkaido study [J]. Sci Total Environ 541:1002–1010
46. Hoffman K, Webster TF, Weisskopf MG et al (2010) Exposure to polyfluoroalkyl chemicals and attention deficit/hyperactivity disorder in U.S. children 12-15 years of age [J]. Environ Health Perspect 118(12):1762–1767

47. Olsen GW, Gillard FD, Burlew MM et al (1998) An epidemiologic investigation of reproductive hormones in men with occupational exposure to perfluorooctanoic acid [J]. J Occup Environ Med 40(7):614–622
48. Toft G, Jonsson BA, Lindh CH et al (2012) Exposure to perfluorinated compounds and human semen quality in arctic and european populations [J]. Hum Reprod 27(8):2532–2540
49. Joensen UN, Bossi R, Leffers H et al (2009) Do perfluoroalkyl compounds impair human semen quality? [J]. Environ Health Perspect 117(6):923–927
50. Raymer JH, Michael LC, Studabaker WB et al (2012) Concentrations of perfluorooctane sulfonate (pfos) and perfluorooctanoate (pfoa) and their associations with human semen quality measurements [J]. Reprod Toxicol 33(4):419–427
51. White SS, Fenton SE, Hines EP (2011) Endocrine disrupting properties of perfluorooctanoic acid [J]. J Steroid Biochem Mol Biol 127(1–2):16–26
52. Chang SC, Ehresman DJ, Bjork JA et al (2009) Gestational and lactational exposure to potassium perfluorooctanesulfonate (k+pfos) in rats: Toxicokinetics, thyroid hormone status, and related gene expression [J]. Reprod Toxicol 27(3–4):387–399
53. Stanifer JW, Stapleton HM, Souma T et al (2018) Perfluorinated chemicals as emerging environmental threats to kidney health a scoping review [J]. Clin J Am Soc Nephrol 13(10):1479–1492
54. Lieder PH, York RG, Hakes DC et al (2009) A two-generation oral gavage reproduction study with potassium perfluorobutanesulfonate (k+pfbs) in Sprague dawley rats [J]. Toxicology 259(1–2):33–45
55. Newsted JL, Beach SA, Gallagher SP et al (2008) Acute and chronic effects of perfluorobutane sulfonate (pfbs) on the mallard and northern bobwhite quail[J]. Arch Environ Contam Toxicol 54(3):535–545
56. Slotkin TA, Mackillop EA, Melnick RL et al (2008) Developmental neurotoxicity of perfluorinated chemicals modeled in vitro [J]. Environ Health Perspect 116(6):716–722
57. Oldham ED, Xie W, Farnoud AM et al (2012) Disruption of phosphatidylcholine monolayers and bilayers by perfluorobutane sulfonate [J]. J Phys Chem B 116(33):9999–10007
58. Chengelis CP, Kirkpatrick JB, Radovsky A et al (2009) A 90-day repeated dose oral (gavage) toxicity study of perfluorohexanoic acid (pfhxa) in rats (with functional observational battery and motor activity determinations) [J]. Reprod Toxicol 27(3–4):342–351
59. Wang ZY, Cousins IT, Scheringer M et al (2015) Hazard assessment of fluorinated alternatives to long-chain perfluoroalkyl acids (pfaas) and their precursors: status quo, ongoing challenges and possible solutions [J]. Environ Int 75:172–179

Environmental Antibiotics: Exposure Monitoring and Health Endpoints

Yingya Zhao and Xirong Xiao

Abstract Antibiotics have long been used to combat human and animal diseases and promote the growth of livestock. Due to the low metabolic rate of antibiotics, they cannot be fully absorbed by human and animal bodies. Most antibiotics are discharged into the environment in the form of active ingredients or metabolites, resulting in residues in the environmental medium. More and more attention has been paid to the environmental pollution caused by unreasonable and irregular use of antibiotics. Antibiotics in the environment could cause different degrees of toxic effects on various environmental media such as water and soil, and even induce the production of resistance genes, which will bring serious threats to human life and health. This chapter introduces the exposure of antibiotics and their effects on ecological environment and human health. Moreover, we focus on the state-of-art researches on antibiotic resistance gene and initiate several prospects of future research.

Keywords Antibiotics · Environment exposure · Health effects · Resistance gene

Abbreviations

Al	aluminum
ANOVA	analysis of variance
ANS	autonomic nervous system

Y. Zhao
School of Public Health, Fudan University, Shanghai, China

Key Laboratory of Public Health Safety, Ministry of Education, Fudan University, Shanghai, China

X. Xiao (✉)
Department of Obstetrics and Gynecology, Obstetrics and Gynecology Hospital of Fudan University, Shanghai, China
e-mail: xiaoxirong@fudan.edu.cn

Y. Zhang (ed.), *Emerging Chemicals and Human Health*,
https://doi.org/10.1007/978-981-32-9535-3_10

As	arsenic
Ba	Barium
BMI	body mass index
BP	blood pressure
Br	bromine
Ca	calcium
Cd	cadmium
Cl	chlorine
Co	Cobalt
Cr	chromium
Cu	copper
DBP	diastolic blood pressure
ED-XRF	energy dispersive X-ray fluorescence
ET-1	endothelin-1
Fe	iron
GWG	gestational weight gain
HDP	hypertensive disorders of pregnancy
ICAM-1	intercellular adhesion molecule 1
IL	interleukin
IQR	interquartile range
K	potassium
MAP	mean arterial pressure
Mg	magnesium
Mn	manganese
Mo	molybdenum
mPEAK	Maternal Psychological and Environmental Assessment of Kids
Na	sodium
Ni	nickel
P	phosphorus
Pb	lead
PM	particulate matter
PTFE	polytetrafluoroethylene
Rb	rubidium
S	sulfur
Sb	antimony
SBP	systolic blood pressure
Se	selenium
Si	silicon
Sn	stannum
Sr	strontium
Ti	titanium
TNFa	tumor necrosis factor alpha
V	vanadium
VCIM-1	vascular cell adhesion molecule 1
Zn	zinc

1 Introduction

Antibiotics, also known as antimicrobials, are secondary metabolites produced by actinomycetes, bacteria, fungi, and other microorganisms at a certain stage of growth or the same and similar substances obtained by direct chemical synthesis or semisynthesis. At low concentrations, they can selectively act on specific segments of bacteria, interfering with metabolism of cells and ultimately inhibiting or killing pathogenic microorganisms [1]. Since the discovery of penicillin by British bacteriologist Fleming in 1928 and the clinical use by Florey and Chain, more than one thousand antibiotics had been developed and utilized, including β-lactam, aminoglycoside, macrolide, tetracycline, quinolone, and sulfonamide. As drugs for the prevention and treatment of diseases in humans and animals, antibiotics had been widely used in the global pharmaceutical, animal husbandry, and breeding industries. Since different drugs show different metabolic states within organisms, 30–90% of the antibiotics ingested by humans and animals are excreted into the environment together with urine and feces instead of being completely absorbed [2, 3]. In addition, along with the rapid development of animal husbandry and aquaculture, the overuse of veterinary antibiotics has greatly increased the excretion and release of antibiotics into the environment.

Antibiotics in the environment are mostly found in the water and the soil. The main sources of antibiotics in the water include hospitals that discharge antibiotics intensively, farms that wash livestock waste, and sewage treatment plants. Antibiotics in the soil mainly come from direct discharge of poultry manure and agricultural fertilization [4, 5]. The large amount of antibiotics residues in the environment not only have toxic effects on microorganisms, plants, and animals but also migrate into human food and drinking water, posing potential risks to human health [6]. Meanwhile, increasing studies showed that long-term abuse of antibiotics could induce resistance genes and gene pollution to the ecological environment, threatening human health seriously [7].

2 Antibiotic Production and Application

In 1999, total consumption of antibiotics in Europe was 13,200 t, among which 8500 t were medical antibiotics and 4700 t were veterinary antibiotics, accounting for 65% and 35% accordingly [8, 9]. A large number of veterinary antibiotics consumption which reached 9300 t in 1999 in the United States turned this country into one of the big consumers of antibiotics [10]. We have changed it to Since the 2000s, the production and application of antibiotics has increased sharply worldwide. In 2002, the global consumption of antibiotics increased from 100,000 t to 200,000 t [11]. In 2003, production of more than 16 antibiotics was more than 7 t in South Korea, among which production of amoxicillin and cephalosporin takes majority with the amount of 96 t and 84 t, respectively [12]. Meanwhile, in the

Czech Republic, application of veterinary antibiotics was also increased to 80 t in 2008 [13]. China is a major producer and consumer of antibiotics, and main types of antibiotics included penicillin, cephalosporin, β-lactamase inhibitors, aminoglycoside, tetracycline, macrolide, quinolone, antituberculosis, and antiviral drugs. Up to 2009, annual production of antibiotics of China had reached 147,000 t, the number of antibiotic varieties had reached 181, and the number of production enterprises had reached 120, with the market size of about 60 billion yuan, ranking the first in the world [14]. However, abuse of antibiotics in China is serious and not negligible. Taking medical antibiotics application as an example, proportion of antibiotics in drug prescriptions in western countries is only 30%, while it is as high as 70% in China [15]. Besides, antibiotics are also widely used in breeding industry to promote the growth of livestock and improve the quality of raw meat. About 50% of antibiotics consumption belongs to breeding industry in China every year, among which tetracycline and oxytetracycline are the main antibiotics [16].

3 Sources of Antibiotics in the Environment

3.1 Medical Antibiotics

Medical antibiotics are one of the main sources of antibiotic pollution in the environment. Antibiotics enter the body mainly by oral administration and injection. After being absorbed by the body, a small part of antibiotics are transferred into inorganic compounds through the body's metabolic reaction, and the antibiotics without metabolic reaction are discharged from the body in the original or active form through the patient's feces and urine. The antibiotics discharged from human body and the residual antibiotics from medical treatment are collected into municipal sewage treatment plants through city and hospital sewers and then discharged into surface water after purification by the sewage treatment plant. However, current sewage treatment technology is not sufficient to completely remove the antibiotics in wastewater, so discharge into surface water is likely to cause subsequent pollution of drinking water sources and soil. Besides, expired antibiotics discarded by hospitals enter landfills and leachate into nearby soil or water layers, leading to soil and groundwater pollution. According to a survey conducted by the World Health Organization, the use rate of antibiotics among hospitalized patients in China is as high as 80%, among which rate of broad-spectrum antibiotics and more than two kinds of antibiotics combined is 58% and the use rate of home-prepared antibiotics has also reached 80%, reflecting the serious abuse of antibiotics in China [17].

3.2 Agricultural Antibiotics

In addition to treating human diseases at therapeutic levels, antibiotics are widely used in animal husbandry and aquaculture at subdose levels to prevent infectious diseases and as antimicrobial growth promoters to accelerate animal growth. Global application amount of antibiotics in animal husbandry and aquaculture is huge, which promotes the development of food industry. Only a small part of antibiotics are metabolized and absorbed after entering livestock and poultry, and about 40–90% of antibiotics are discharged with animal urine and feces in the form of original drugs or metabolites into the environment [18]. Veterinary antibiotics are commonly administered in fields along with animal feces and urine. Some of them seep directly into underground water, while others enter the soil, rivers, or lakes after being washed by rain, causing water and soil pollution. About 70–80% of antibiotics applied in aquaculture end up in the water environment, with some going directly into the water and others being trapped in the silt. Studies showed that antibiotics were not absorbed by aquatic organisms but excreted into the water through excreta and remained in the water body with large quantities or settled in the mud land in aquaculture farms. Antibiotic residues in water bodies are discharged into surface runoff which will pollute surface water, marine, and other water environments. Antibiotics in the bottom silt generally enter the soil through fertilization and then enter the water environment by rainwater erosion.

3.3 Antibiotics in Pharmaceutical Wastewater

There are various drugs and by-products left in the wastewater produced by pharmaceutical factories, which are characterized by high concentrations of organic compounds and resistance to degradation, and there are still drugs left or adsorbed in treated wastewater and sludge. In some factories, large amounts of antibiotics and their oxygenated products remain in sewage. Antibiotic preparation includes fermentation, chemical synthesis, extraction, and drug preparation, when wastewater containing several biotoxic substances and active antibiotics that are difficult to be degraded is discharged. These high concentrations of biotoxic substances and antibiotics inhibit the proliferation of microorganisms in wastewater biochemical treatment, resulting in more difficult degradation of antibiotics in wastewater [19]. If pharmaceutical factories do not treat wastewater according to relevant standards and discharge in violation of laws and regulations, the antibiotic residues will transfer into surface water and other water environment, which could seriously pollute the ecological environment. Many pharmaceutical factories still use traditional sewage treatment methods which are inefficient to degrade antibiotics in wastewater completely. The illegal discharge of pharmaceutical wastewater is a major cause of antibiotic pollution in environmental water [20].

4 Antibiotic Concentrations in the Environment

Antibiotics have been widely found in various environments including water, soil, animal, and plant samples with high species diversity. In the water environment, antibiotics are widely detected in surface water and groundwater (even hundreds of meters deep). Sacher et al. collected 108 underground water samples from Germany and detected 60 kinds of antibiotics at levels of μg/L [21]. In 2009, Tong et al. detected several typical quinolone antibiotics in groundwater and surface water in Hubei province of China, and the concentration of quinolones in surface water (0.01 g/L) was significantly higher than that in groundwater (0.001 g/L) [22]. The study of Sun et al. on the quality of surface water in Xiamen city of China also showed that the surface water was polluted by antibiotics [23]. In addition to the presence of antibiotics in surface water in areas with high levels of human activity, antibiotic contamination had also been found in surface water in some background areas. Golet et al. detected quinolones at levels of ng/L in streams in the watershed of the Glatt Valley in Switzerland [24].

In addition to the water environment, antibiotics are also widely detected in feces, soil, and crops. Ma et al. found that the total concentration of antibiotics in pig manure was as high as 10.83 mg/kg [25]. Zhang et al. also detected high levels of antibiotics (1.57–6.29 mg/kg) in chicken manure, cow manure, and pig manure and found that soil was seriously and persistently contaminated with antibiotics after the application of sludge or livestock manure containing antibiotics [26]. A variety of antibiotics including sulfa, tetracycline, and macrolide were detected in soil in the United Kingdom and Canada [27]. A study by Zhou et al. on farmland soil in Northern China showed that the concentration of antibiotics in ordinary soil was at the level of g/kg, while the concentration of antibiotics in soil applied with organic fertilizers was higher, with a wider range, fluctuating between the level of g/kg and mg/kg, among which the concentration of sulfadimethylpyrimidine was as high as 900 mg/kg [28]. Tai found that quinolones, tetracycline, and sulfonamides were detected in the soil in the Pearl River delta region of China and even high-grade vegetable bases (such as pollution-free vegetable production base, green vegetable production base, and organic vegetable production base) at the level of dozens to hundreds of mg/kg [29].

Antibiotics migrate into farmland soil system through agricultural sludge, organic fertilizers, and irrigation water, causing antibiotic pollution in the soil, leading to absorption and accumulation of antibiotics in vegetables, and then threatening human health through the food chain.

Different plant varieties can accumulate different amounts of antibiotics in soil, and the accumulation capability for antibiotics increases with higher antibiotic concentrations in soil. Wu's study on quinolone antibiotics in vegetables in Guangzhou and Dongguan of China showed that the detection rates of four quinolone antibiotic compounds (ciprofloxacin, enoxacin, norfloxacin, and lomefloxacin) were more than 90%, and the average total content (dry weight) was 74.38 g/kg (Guangzhou) and 46.34 g/kg (Dongguan), respectively; the highest antibiotic content (mean:

91.29 g/kg) was found in fruits and melons, and the lowest antibiotic content was found in leafy vegetables (mean: 52.30 g/kg) [30]. Lillenberg et al. found that cucumber, barley, and lettuce all absorbed and accumulated ciprofloxacin and enoxacin [31].

5 The Harm of Antibiotic Residues in the Environment

5.1 Toxic Effects of Antibiotics on Aquatic Environment

5.1.1 Effects on Microflora in Water

High concentrations of antibiotics inhibit or kill microorganisms, but the concentration of antibiotics in the water environment is generally too low to have a strong killing effect on microorganisms. However, when the microbial flora in the water environment is exposed to low-dose antibiotics for a long time, it is likely to lead to the emergence of drug-resistant strains. Drug-resistant strains transfer drug-resistant genes to the next generation or to other organisms through gene mutation. Under natural selection, after the proliferation of drug-resistant strains, antibiotics act on higher organisms through the food chain, disrupting the balance of the ecosystem [32]. A large number of studies showed that antibiotics induce drug resistance in pathogenic bacteria, and drug-resistant strains in the human body accelerate the degradation of drugs, reducing the inhibitory effect of drugs on bacteria and weakening the effect of drugs. Bacteria with resistance to one or more antibiotic drugs can be detected in sewage from treatment plants. If the sewage is not treated properly, these bacteria are likely to flow into surface water through fertilization and irrigation, or enter human body through the food chain, which will cause danger to human health ultimately. Yu et al. showed that in Wenyu river of China, about 45% of *Escherichia coli* had single-drug resistance or multiple-drug resistance, and the drug-resistant bacteria had the strongest resistance to tetracycline, sulfa, and ampicillin [33].

5.1.2 Effects on Aquatic Life

Antibiotics have toxic effects not only on aquatic microorganisms but also on other nontarget organisms in the water environment. The generally low concentration of drugs in the water environment does not cause acute poisoning to aquatic life, but long-term and low-concentration drug exposures generate chronic toxic effects [34]. Large amounts of antibiotic residues in the environment or in organisms can cause certain harm to aquatic organisms with similar or identical tissues or cells. Different antibiotics have different toxicity to aquatic organisms. A previous study found that erythromycin, ciprofloxacin, or sulfamethoxazole inhibited the photolytic reaction of the crescent algal, among which erythromycin had a stronger inhibitory effect on the crescent algal than the other two. Moreover, oxidative products of some drugs are more toxic to some target organisms than the original compounds. For example, amoxicillin degradation products are more toxic to macroalgae than amoxicillin itself [35]. Jiang et al. tested the toxicity of tetracycline, sulfonamide, and quinolone with luminescent bacteria, and the results showed that the toxicity of these antibiotics was not significant, but the difference in toxicity was reflected in EC50 (36.39–938.42 mg/L) [36]. In general, except for the various types of antibiotics, they also have different and complex degradation products. Studies on the combined toxic effects of the mother and the degradation products on aquatic organisms are still in the initial stage, and more toxic effects still need to be further studied.

5.2 *Toxic Effects of Antibiotics on Soil Environment*

5.2.1 Effects on Soil Microorganisms

Soil microorganisms are an important part of soil environment and the power source of material transformation in soil. When antibiotics enter soil through fecal discharge or sewage irrigation, they will be adsorbed by soil particles and accumulate for a long time. When antibiotics in soil accumulate to a certain amount, they will have effects on bacteria, fungi, actinomycetes, and other organisms in soil significantly, not only affecting the community structure of microbial species but also inducing microbial resistance and inhibiting or enhancing microbial metabolic pathways. Hammesfahr et al. added sulfadiazine at different concentrations into the soil for experiment, and the results showed that sulfadiazine reduced the amount of phospholipid fatty acids and the proportion of bacteria and fungi in the soil and even changed the microbial community structure after 2 months [37]. Liu et al. used the biolog-eco technology to analyze the structural diversity of soil microorganisms under different penicillin concentrations and found that when the concentration of penicillin was 800 mg/kg, the soil microbial diversity index was significantly reduced [38]. Moreover, antibiotics can destroy the microecological structure in soil, thus affecting the decay and decomposition of organic matters and reducing soil fertility. Dijck et al. showed that antibiotics in livestock and poultry feces inhibited the growth of a variety of microorganisms, including bacteria, fungi, and actinomycetes. The reduction of these microorganisms hindered the nitrification and mineralization in soil, impeding the normal nutrient circulation of soil [39].

5.2.2 Effects on Soil Flora and Fauna

In nature, the roots of most plants are fixed in soil, which is the substrate for the growth and development of plants. They can provide nutrients, water, and air for plants. When antibiotics enter the soil, plants will produce obvious enrichment.

Migliore et al. obtained through experiments that the roots of barley and corn contained sulfadimethoxypyrimidine added in the culture soil [40]. Kumar et al. also demonstrated that maize, onions, and kale absorbed aureomycin from soil [41]. Bao et al. found that enoxacin and oxytetracycline in soil accumulated in water celery through a soil culture experiment, and the accumulation order of antibiotics in each organ of water celery was root > leaf > stem [42]. Antibiotics can affect the growth and development of plants after entering the body. Similar to most pollutants, high concentrations of antibiotics inhibit the growth of plants, while low concentrations of antibiotics promote the growth of plants [43]. Kong et al. found that oxytetracycline at a concentration higher than 0.002 mmol/L significantly inhibited the growth of alfalfa. The inhibition rate of 0.002–0.2 mmol/L oxytetracycline on alfalfa stem and root reached 61% and 85%, respectively [44]. Zhang et al. studied the effects of the addition of different doses of tetracycline on the growth and development of flue-cured tobacco by pot experiment system, and the results showed that the addition of a low amount of tetracycline promoted the increase of tobacco plant height, and the addition of medium and high amounts of tetracycline significantly inhibited the plant height [45].

Soil animals are important consumers of the soil environment and play an important role in the ecosystem. They not only assimilate all kinds of beneficial substances in soil to build themselves, but also return excreta to soil, thus continuously transforming the environment. There are few studies on the toxicity of antibiotics to soil animals. Dong et al. studied the genetic toxicity of tetracycline and aureomycin to earthworm by measuring the damage of earthworm DNA. They found that the genotoxicity of the two antibiotics at the concentration of 0.3–300 mg/kg had a significant dose-effect relationship, and the genotoxicity of aureomycin was greater in the short term [46]. Gao et al. studied the avoidance behavior of *Eisenia fetida* earthworm in soil containing oxytetracycline. After 48 h of oxytetracycline exposure, they found that no earthworm was dead within the range of 0–2560 mg/kg [47]. It can be seen that the toxicity of antibiotics to soil animals is weak, so under natural conditions, antibiotics in soil have almost no direct impact on soil animals, but its indirect impact on the ecosystem still cannot be ignored.

5.3 Effects of Antibiotic Resistance Genes (ARGs)

Another important threat to ecosystems from antibiotics is the induction of resistance genes. In recent years, with the frequent use of antibiotics in the pharmaceutical industry, animal husbandry, and aquaculture, antibiotic resistance has increased significantly in diversity, intensity, and transmission speed, and a variety of drug-resistant bacteria are emerging one after another.

ARGs are genetic factors for acquiring resistance to antibiotics. Since Pruden et al. firstly identified antibiotic resistance gene as a new pollutant, more and more attention has been paid to the problem [48]. To date, ARGs have been detected in surface waters and soils in many countries. Isabel et al. studied the distribution of

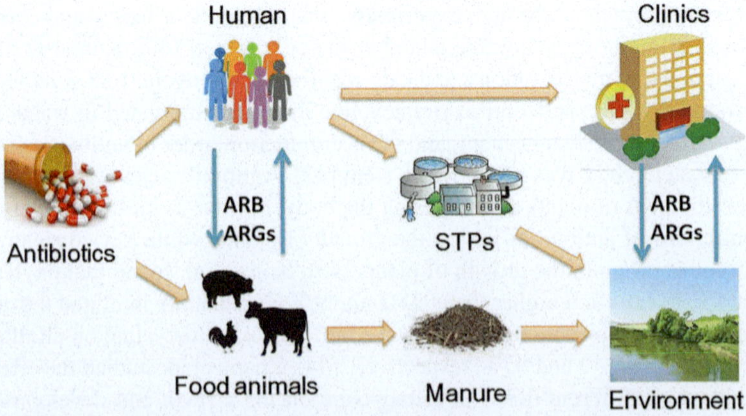

Fig. 1 Antibiotic resistance in the environment and its potential health impacts

ARGs in the Avero River in Portugal and found that 77.8% of Enterobacteria and 10.5% of *Aeromonas* contained at least one lactamase gene among the bacteria resistant to ampicillin in water [49]. Hsu et al. studied the distribution of ARGs in surface water and soil near the pig farm in Taipei. The results showed that both surface water and soil contained sulfonamide resistance genes *sul1* and *sul2*, and the soil also contained sulfonamide resistance genes *sul3* [50]. Zhang et al. tested the species and abundance of ARGs in the soil of 11 vegetable fields in Beijing, and the results showed that the detection rates of sulfa resistance genes *sul1*, *sul2* and tetracycline resistance genes *tetL* in the soil were all 100% [51]. The above research shows that ARGs pollution in the environment has become increasingly serious, and the law of ARGs transfer and extinction is still not clear, so it is necessary to take actions as soon as possible.

The World Health Organization (WTO) claimed that antibiotic resistance is one of the important public health problems in the twenty-first century [52]. In the environment, ARGs are spread by vertical and horizontal transfer. On the one hand, ARGs are transferred to the offspring bacteria by means of vertical transfer, which leads to the continuous increase of resistant bacteria, changing the microbial structure and community of the ecosystem, and damaging the stability and balance of the ecosystem. On the other hand, ARGs are transferred horizontally into other organisms and the food chain. Once these bacteria enter the human body, they will bring a great difficulty to the clinical treatment of human diseases and a seriously danger to the public (Fig. 1).

5.4 The Harm of Antibiotics to Human Health

Antibiotics in the environment, if not controlled, will accumulate slowly in different media and eventually harm human health. Livestock animals have long been consuming and drinking feed and water containing added antibiotics, which can be absorbed in small amounts by the animal intestinal tract and transported with the blood to various organs. Antibiotic residues are found not only in animal tissues but also in by-products such as meat, eggs, and milk. Manure or water containing antibiotics is applied to vegetables, rice, and other agricultural products for a long time, and antibiotics also exist in their food products. When human eat animal and plant products containing antibiotic residues, through the principle of biological enrichment, antibiotics in animal and plant food are transmitted to people along the food chain, causing allergic reactions in the population and food poisoning in severe cases. Some drugs even have carcinogenic, teratogenic, and mutagenic or hormone effects which seriously interfere with human physiological functions [53]. Moreover, the existing water purification processes lack the removal measures for antibiotics and corresponding detection device. The antibiotic residues in the water environment can directly enter the human body through drinking water. The content of antibiotics in drinking water is generally low, and the direct acute allergic reactions or toxic effects are not obvious. However, antibiotics can accumulate in the human body through biological enrichment, thus increasing the drug resistance of bacteria in the body, reducing the function of the immune system, and interfering with the normal function of the endocrine system [54].

Improper use of antibiotics in itself has certain harm to human health. β-Lactam antibiotics such as penicillin can cause allergic reactions in the human body; gentamicin is very toxic to the kidney; quinolones can increase the body's sensitivity to light; and tetracycline can seriously affect the development of children's teeth [55]. In addition, some important organs are not fully developed during childhood, and quinolones may have adverse effects on the development of bones, liver, kidneys, and so on. The guidelines for clinical use of antimicrobial drugs stipulate that such drugs should be avoided for minors under the age of 18 years, [56] which reflects from another perspective that the potential threat of antibiotic residues in the environment to human health should arouse more public attention.

6 Conclusion

To sum up, although the concentration of antibiotics in the environment is low, its harm to the ecological environment and human health is very great. In 1990, European Union countries started risk assessments on antibiotics in the environment. The United States also started investigations and research on antibiotic drug use and environmental residue levels in 2001 [57]. Due to the rapid development of China's medical and health undertakings and pharmaceutical industry, the production and use of antibiotics are extremely extensive. However, the abuse of antibiotics is also very serious. For example, some farmers use antibiotics in pursuit of excessive profits, resulting in excessive residues of antibiotics in breeding products. In addition, at present, China's sewage treatment technology requires further promotion, and there are relatively few reports on antibiotic pollution in the environment. The antibiotic pollution in China calls for urgent countermeasures. We are challenged with many problems, such as regulations of antibiotic degradation, the controls of resistance gene migration and extinction, the studies of toxic mechanisms of antibiotics, and the assessment of impacts of antibiotics on human living environment and ecosystem. Therefore, it is necessary to establish an environmental and ecological risk assessment system, optimize the detection methods with high sensitivity and accuracy, strengthen existing conventional processing techniques, explore new techniques for efficient processing, establish standards for antibiotics in various environmental media, and revise relevant laws and regulations, so as to control the pollution of antibiotics in the environment and realize the reuse of resources.

References

1. Brandt KK, Amézquita A, Backhaus T et al (2015) Ecotoxicological assessment of antibiotics: a call for improved consideration of microorganisms [J]. Environ Int 85:189–205
2. Halling Sorensen B, Nielsen SN, Lanzky PF et al (1998) Occurrence, fate and effects of pharmaceutical substances in the environment: a review [J]. Chemosphere 36(2):357–394
3. Puckowski A, Mioduszewska K, Lukaszewicz P et al (2016) Bioaccumulation and analytics of pharmaceutical residues in the environment: a review [J]. J Pharm Biomed Anal 127:232–255
4. Chee Sanford JC, Mackie RI, Koike S et al (2009) Fate and transport of antibiotic residues and antibiotic resistance genes following land application of manure waste [J]. J Environ Qual 38(3):1086–1108
5. Xu Yonggang, Yu Wantai, Ma Qiang et al (2015) The antibiotic in environment and its ecotoxicity: a review [J]. Asian J Ecotoxicol 3:11–27
6. Du LF, Liu WK (2012) Occurrence, fate and ecotoxicity of antibiotics in agro-ecosystems: a review [J]. Agron Sustain Dev 32(2):309–327
7. Wellington EMH, Boxall ABA, Cross P et al (2013) The role of the natural environment in the emergence of antibiotic resistance in Gram-negative bacteria [J]. Lancet Infect Dis 13(2):155–165
8. Martínez Carballo E, González Barreiro C, Scharf S et al (2007) Environmental monitoring study of selected veterinary antibiotics in animal manure and soils in Austria [J]. Environ Pollut 148(2):570–579
9. Carvalho IT, Santos L (2016) Antibiotics in the aquatic environments: a review of the European scenario [J]. Environ Int 94:736–757
10. Sarmah AK, Meyer MT, Boxall ABA (2006) A global perspective on the use, sales, exposure pathways, occurrence, fate and effects of veterinary antibiotics (VAs) in the environment [J]. Chemosphere 65(5):725–759
11. Kümmerer K (2009) Antibiotics in the aquatic environment: a review [J]. Chemosphere 75(4):417–434
12. Lee YJ, Lee SE, Lee DS et al (2008) Risk assessment of human antibiotics in Korean aquatic environment [J]. Environ Toxicol Pharmacol 26(2):216–221
13. Zounkova R, Kliemesova Z, Nepejchalova L et al (2011) Complex evaluation of ecotoxicity and genotoxicity of antimicrobials oxytetracycline and flumequine used in aquaculture [J]. Environ Toxicol Chem 30(5):1184–1189
14. Zhang zhenghai. Three major changes of antibiotics in China in the past 10 years, with the output and export volume ranking first in the world [N]. Pharm News. 2010-02-02
15. Bruce JR, Paul KSL, Michael M (2005) Emerging chemicals of concern: Pharmaceuticals and personal care products (PPCPs) in Asia, with particular reference to Southern China [J]. Mar Pollut Bull 50:913–920

16. Yu jie, hong tao. The glory and hardship in the era of antibiotics [N]. China Educ Daily. 2003-09-12
17. People's Daily online. The misuse of antibiotics may cause the virus mutation [R/OL]. [2003-06-15]. http://www.people.com.cn/GB/paper1787/9423/872497.html. Accessed 5 May 2019
18. Wu X, Wei Y, Zheng J et al (2011) The behavior of tetracycline and their degradation products during swine manure composting [J]. Bioresour Technol 102(10):5924–5931
19. Ye BX, Zhang L (2015) Analysis of the pollution status and health risk of antibiotics in water environment and drinking water [J]. Int J Environ Res Public Health 32(2):173–178
20. Chen T (2010) Research on a variety of antibiotics of wastewater treatment plants in Guangzhou and Zhuhai by HPLC/LC-MS analysis method [D]. Jinan University, Guangzhou
21. Sacher F, Lange FT, Brauch H (2001) Pharmaceuticals in groundwater: analytical methods and result of a monitoring program in Baden Wurttemberg, Germany [J]. J Chromatogr A 938(1):199–210
22. Tong L, Li P, Wang YX et al (2009) Analysis of veterinary antibiotic residues in swine wastewater and environmental water samples using optimized SPELC/MS/MS [J]. Chemosphere 74(8):1090–1097
23. Sun GD, Su ZY, Chen M et al (2009) Simultaneous determination of tetracycline and quinolone antibiotics in environmental water samples using solid phase extraction-ultra pressure liquid chromatography coupled with tandem mass spectrometry [J]. Chin J Chromatogr 27(4):54–58
24. Golet EM, Alder AC, Hartmann A et al (2001) Trace determination of fluoroquinolone antibacterial agents in solid-phase extraction urban wastewater by and liquid chromatography with fluorescence detection [J]. Anal Chem 73(15):3632–3638
25. Ma Y, Chen ZZ, Zeng ZL (2007) Effects of enoxacin on soil microbial community functional diversity [J]. Acta Ecol Sin 27(8):3400–3406
26. Zhang JQ, Dong YH, An Q et al (2005) Behavior of veterinary antibiotics in soil environment [J]. Soil 37(4):353–361
27. Aust MO, Godlinski F, Travis GR et al (2008) Distribution of sulfamethazine, chlortetracycline and tylosin in manure and soil of Canadian feedlots after sub therapeutic use in cattle [J]. Environ Pollut 156(3):1243–1251
28. Zhou QX, Luo Y, Wang ME (2007) Environmental residues, ecotoxicity and resistance gene contamination of antibiotics [J]. J Ecotoxicol 2(3):243–251
29. Tai Yiping (2010) Study on the characteristics of typical antibiotic contamination in soil of vegetable bases in the Pearl River delta region [D]. Jinan university, Guangzhou
30. Wu Xiaolian (2011) Contamination characteristics and health risks of quinolones antibiotics in vegetables in the Pearl River delta region [D]. Jinan University, Guangzhou
31. Lillenberg M, Litvin SV, Nei L et al (2010) Enrofloxacin and ciprofloxacin uptake by plants from soil [J]. Agron Res 8(1):807–814
32. Richardson BJ, Lam PKS, Martin M (2005) Emerging chemicals of concern: pharmaceuticals and personal care products (PPCPs) in Asia, with particular reference to southern China [J]. Mar Pollut Bull 50:913–920
33. Yu Daojin, Zeng Zhenling, Chen Jaoshi (2005) Effects of oxytetracycline residues on the resistance of sediment bacteria [C]. In: Proceedings of the fourth national conference of the Chinese society of toxicology. Chinese Society of Toxicology, Shenyang, p 1
34. Wang Penghua, Yuan tao, Tan youming (2008) Effects of drug pollution on aquatic organisms and human health [J]. J Environ Health 25(2):172–174
35. Gbel A, Thomsen A, McArdell CS et al (2005) Occurrence and sorption behavior of sulfonamides, macrolides, and trimethoprim in activated sludge treatment [J]. Environ Sci Technol 39(11):3981–3989
36. Jiang L, Lin Z, Hu X et al (2010) Toxicity prediction of antibiotics on luminescent bacteria, *Photobacterium phosphoreum*, based on their quantitative structure-activity relationship models [J]. Bull Environ Contam Toxicol 85(6):550–555

37. Hammesfahr U, Heuer H, Manzke B et al (2008) Impact of the antibiotic sulfadiazine and pig manure on the microbial community structure in agricultural soils [J]. Soil Biol Biochem 40(7):1583–1591
38. Liu Yan, Zhang Zhenhua, Bao Wangbo et al (2016) Effect of penicillin on soil microbial community structure [J]. J Ecol Rural Environ 2:309–314
39. Van Dijck P, vande Voorde H (1976) Sensitivity of environmental microorganisms to antimicrobial agents [J]. Appl Environ Microbiol 31(3):332–336
40. Migliore L, Civitareale C, Cozzolino S et al (1998) Laboratory models to evaluate phytotoxicity of sulphadimethoxine on terrestrial plants [J]. Chemosphere 37(14/15):2957–2961
41. Kumar K, Gupta SC, Baidoo SK et al (2005) Antibiotic uptake by plants from soil fertilized with animal manure [J]. J Environ Qual 34(6):2082–2085
42. Bao Chenyan, Gu Guoping, Zhang Mingkui (2016) Effect of veterinary antibiotics stress on growth and antibiotics accumulation of *Oenanthe javanica* DC [J]. Chin J Soil Sci 1:164–172
43. Kong Wei Dong, Zhu Yong Guan (2007) A review on ecotoxicology of veterinary pharmaceuticals to plants and soil microbes [J]. Asian J Ecotoxicol 1:1–9
44. Kong WD, Zhu YG, Liang YC et al (2007) Uptake of oxytetracycline and its phytotoxicity to alfalfa [J]. Environ Pollut 147(1):187–193
45. Zhang Jixu, Shen Guoming, Kong Fanyu et al (2017) Effect of tetracyline on the growth and photosynthesis of flue-cured tobacco [J]. J Agro Environ Sci 1:48–56
46. Dong L, Gao J, Xie X et al (2012) DNA damage and biochemical toxicity of antibiotics in soil on the earthworm Eisenia fetida [J]. Chemosphere 89(1):44–51
47. Gao M, Lyu M, Han M et al (2016) Avoidance behavior of Eisenia fetida in oxytetracycline and heavy metal-contaminated soils [J]. Environ Toxicol Pharmacol 47:119–123
48. Pruden APRT, Storteboom H, Carlson KH (2006) Antibiotic resistance genes as emerging contaminants: studies in Northern Colorado [J]. Environ Sci Technol 40:7445–7450
49. Henriques IS, Fonseca F, Alves A et al (2006) Occurrence and diversity of integrons and β-lactamase genes among ampicillin-resistant isolates from estuarine waters [J]. Res Microbiol 157(10):938–947
50. Hsu JT, Chen CY, Young CW et al (2014) Prevalence of sulfonamide-resistant bacteria, resistance genes and integron-associated horizontal gene transfer in natural water bodies and soils adjacent to a swine feedlot in northern Taiwan [J]. J Hazard Mater 277:34–43
51. Zhang Lanhe, Wang jiajia, Ha xuejiao et al (2016) Distribution characteristics of antibiotic resistance genes in vegetable field soil in Beijing [J]. Environ Sci 11:4395–4401
52. Cizmas L, Sharma VK, Gray CM et al (2015) Pharmaceuticals and personal care products in waters: occurrence, toxicity and risk [J]. Environ Chem Lett 13(4):381–394
53. Chen qiuying, Jin caixia, Lu shanhua et al (2008) Research progress on veterinary drug residues and their impact on ecological environment [J]. Anhui Agric Sci 36(16):6943–6952
54. Lu mei, Gao xiang, Li Shuwen (2008) Determination of tetracycline antibiotic residues in aquatic products by high performance liquid chromatography [J]. Environ Monit Manage Technol 20(5):38–39
55. Kümmerer K (2009) Antibiotics in the aquatic environment. A review part I [J]. Chemosphere 75(4):417–434
56. General office of the ministry of health. Notice about on the management of clinical application of antibacterial drugs [EB/OL]. http://www.moh.gov.cn/mohbgt/s9508/200903/39723. Accessed 5 May 2019
57. Benbrook CM (2002) Antibiotic drug use in U.S. aquaculture [EB/OL]. http://www.iatp.org/library/antibiotics. Accessed 5 May 2019

Correction to: Exposure Assessment of Emerging Chemicals and Novel Screening Strategies

Qingyang Zhu and Haixia Dai

Correction to:
Chapter 2 in: Y. Zhang (ed.), *Emerging Chemicals and Human Health*, https://doi.org/10.1007/978-981-32-9535-3_2

The original version of the book was published without the following Acknowledgement. Chapter no 2, **Acknowledgment** This work was modified from the paper published in *Environmental Health Perspectives* [50] and *Environment International* [51]. The related contents are reused with permission.

References

50. Dennis KK, Marder E, Balshaw DM, Cui Y, Lynes MA, Patti GJ, Rappaport SM, Shaughnessy DT, Vrijheid M, Barr DB (2017) Biomonitoring in the era of the exposome[J]. Environ Health Perspect 125:502–510
51. Johns LE, Cooper GS, Galizia A, Meeker JD (2015) Exposure assessment issues in epidemiology studies of phthalates[J]. Environ Int 85:27–39

The updated version of the chapter can be found at
https://doi.org/10.1007/978-981-32-9535-3_2

Y. Zhang (ed.), *Emerging Chemicals and Human Health*,
https://doi.org/10.1007/978-981-32-9535-3_11

The manufacturer's authorised representative in the EU is Springer Nature Customer Service Centre GmbH, Europaplatz 3, 69115 Heidelberg, Germany. If you have any concerns regarding our products, please contact ProductSafety@springernature.com

Printed and bound by CPI Group (UK) Ltd, Croydon, CR0 4YY
30/07/2026
02182086-0004